Study Guide to Accompany
Integrative Statistics
for the Social &
Behavioral Sciences

Prepared by

Renee L. Ha • James C. Ha
University of Washington

Ashley C. Maliken

Los Angeles | London | New Delhi
Singapore | Washington DC

For information:

SAGE Publications, Inc.
2455 Teller Road
Thousand Oaks, California 91320
E-mail: order@sagepub.com

SAGE Publications Ltd.
1 Oliver's Yard
55 City Road
London EC1Y 1SP
United Kingdom

SAGE Publications India Pvt. Ltd.
B 1/I 1 Mohan Cooperative
    Industrial Area
Mathura Road, New Delhi 110 044
India

SAGE Publications Asia-Pacific
    Pte. Ltd.
33 Pekin Street #02-01
Far East Square
Singapore 048763

ISBN 9781452205250

11 12 13 14 15 10 9 8 7 6 5 4 3 2 1

Acquisitions Editor:	Vicki Knight
Associate Editor:	Lauren Habib
Editorial Assistant:	Kalie Koschielak

Chapter One

Study Guide

Notes to Students

As you begin your course in statistics, it is likely that you are feeling some anxiety about the material. It is common for students to hear "horror stories" about the course and to be concerned about their performance in the class. First of all, we'd like to remind you that the mathematics in this textbook are relatively simple and consist of addition, subtraction, division, multiplication, and squaring. Even if you do not feel that your mathematical ability is your strongpoint, you should be able to handle the math in this course. We have a few suggestions to aid you in tackling the material.

The most important tip that we can offer you in this course is to stay on top of the material. This is a course that relies on the chapters that have come before it. In fact, Chapters 1–6 are the fundamentals of statistics. It isn't until the end of Chapter 6 that you really start doing many statistical tests, but once you do you will rely on the material that came before that chapter. If you were weak on the concepts in a particular chapter (say, Chapter 5), it can hurt your performance on later material. Students who fall behind and try to catch up have the most difficulty with the course materials and are probably some of the students generating the "horror stories" about statistics. This is one course that requires you to come to class and read the textbook. A few students might be able to do one or the other, but most students need to do both.

It's probably the conceptual material in statistics that challenges the most students, not the math. If you find that you are having difficulty understanding the concepts of probability or sampling distributions, you are probably not alone. These concepts take some time to sink in for most of us, and asking questions and rereading the chapters can be a big help.

Our last tip would be to work as many problems as you can. It's common for students to work just a few problems, but working more problems tends to make you more accurate and proficient on examinations. Remember that you'll have more time on the test if you can "blow through" the math because you've worked so many problems prior to the test. If you've only worked a few problems and you are anxious during an exam, it can take a long time to work each problem and the clock (and statistics) can become your enemy. Working extra problems from the exercises and the study guide will give you speed, accuracy, and confidence when you take your exams. Do these extra problems even if you do not get credit for them in your course. If you need even more problems to work, find other statistics textbooks in the library or ask your instructor for assistance.

Chapter Two

Study Guide Questions

Notes to Students

This chapter is a perfect example of the cumulative nature of the material in this course. The terms and notation that you'll learn in this chapter will appear in almost every chapter of this book, including the last chapter! It's critical to learn these terms or the concepts that come up later will *really* seem like gibberish to you!

The most common error students make in learning the notation is to confuse the "Sum of the Squared X's" with the "Sum of the X's, Squared". Pretty confusing, huh? Remember that you can dispense with that confusion by carefully noting where the squaring takes place. If the notation is simplified, as it often is in practical use, then it is a matter of differentiating between $\sum X^2$ and $\left(\sum X\right)^2$. The first symbol tells you to sum the squared scores and the second symbol tells you to square the sum of the scores! Working a few of these problems will also aid you in distinguishing between the two forms of notation.

Make sure that the differences between the measurement scores are clear in your mind, as you'll actually choose different statistical tests depending on which measurement scale is in a story problem. You may need to come back and review this material later in the course as you are faced with choosing the appropriate statistical test.

Conceptual Questions

Multiple Choice Questions

1. Which of the following variables has been labeled with an incorrect measuring scale?
   (a) The number of students in a psychology class: ratio
   (b) Ranking of runners in a marathon: ordinal
   (c) Finishing order in a poetry contest: ordinal
   (d) Self-rating of anxiety level by students in a statistics class: ratio

2. _____ may include a measure of central tendency and shape of the distribution, while _____are used to determine characteristics of a population to test hypotheses.
   (a) Inferential statistics, descriptive statistics
   (b) Descriptive statistics, inferential statistics
   (c) Population statistics, sample statistics
   (d) Variance statistics, inferential statistics

3. $\sum X^2$ is read as "_____" while $\left(\sum X\right)^2$ is read as "_____".

 (a) mean score of X, squared; mean score of squared X's
 (b) sum of X's, squared; sum of squared X's
 (c) sum of squared X's; sum of X's, squared
 (d) mean score of squared X's; mean score of X's, squared

Short Answer Questions

Identify the scale (nominal, ordinal, interval, or ratio) for each of the following variables:

 4. Number of bicycles ridden by graduate students to campus

 5. Type (i.e., brand) of bicycles ridden by graduate students to campus

 6. The IQ of the professors in the psychology department (*assume scaled standard scores*)

 7. Knowledge of physics rated on a 6-point Likert scale ranging from *very poor* (1) to *excellent* (6)

 8. The weights of a group of supermodels

Calculation Questions

Given the data $X_1 = 1$, $X_2 = 4$, $X_3 = 5$, $X_4 = 8$, $X_5 = 10$, calculate the following:

 9. $\sum X^2$

 10. $\sum (X+5)$

 11. $\sum X+5$

 12. $\sum (X^2 - 3)$

 13. $\sum X^2 - 3$

 14. $\left(\sum X\right)^2$

 15. $1/\left(\sum X\right)$

A researcher is interested in studying weight of adolescent females ages 10–13 to determine if weights in 2008 have significantly increased since 1988. She measures the height of a randomly selected sample of women. Assume the heights are normally distributed.

Weight (lbs)		
100	103	93
90	91	101
85	89	79
	87	

16. What is the N?

17. What is the mean weight?

18. What is the range of scores?

19. What is the Sums of Squares?

20. What type of data are these values?

Chapter Two Study Guide Answers

1. d

2. b

3. c

4. ratio

5. nominal

6. interval

7. ordinal

8. ratio

9. 206

10. 53

11. 33

12. 191

13. 203

14. 784

15. .0357

16. 10

17. 91.8lbs

18. 24

19. $\Sigma(X_i - X)^2 = 523.6$

20. interval

Chapter Three

Study Guide Questions

Notes to Students

This chapter covers frequency distributions that are presented in table and graphic displays, as well as a number of other graphs used to display continuous and discrete variables. In this chapter we've used the terms "average" and "center" to refer to the middle score. The graphs in this chapter and the descriptive statistics we'll address in the next chapter are used to provide a summary of data, and many of them are incorporated into determining whether certain assumptions of inferential tests have been met. This material is laying the groundwork for the more sophisticated statistical analyses that you'll be doing later in the course. It should also help you in reading and understanding the graphs you see in published manuscripts. Attention to detail is important in performing the frequency table problems successfully.

Conceptual Questions

Multiple Choice Questions

1. Scatterplots are an inappropriate way to represent _____ data.
    (a) interval
    (b) ratio
    (c) continuous
    (d) nominal

2. If a distribution is described as positively skewed, that implies:
    (a) It is normal
    (b) More data points occur at the lower end of the x-axis
    (c) More data points occur at the higher end of the x-axis
    (d) The data are clustered around the center of the distribution

3. When creating a graph, the x-axis represents the _____ while the y-axis represents the _____.
    (a) dependent variable, independent variable
    (b) independent variable, dependent variable
    (c) range of scores, mean
    (d) mean, range of scores

Short Answer Questions

4. If you were to create a histogram with age on the x-axis and number of medications taken daily on the y-axis, and your sample were Washington State citizens, describe how this distribution may look.

5. For each of the following examples, determine which type of graph (bar graph, histogram, scatterplot) would be the most appropriate representation:

     (a) Hair color of students in a classroom

     (b) Age of first employment

     (c) Miles commuted per week

     (d) Dollars earned based and years of education

Calculation Questions

6. Convert the following raw scores into an appropriate frequency distribution using $i = 100$ and complete the table below:

80	123	430	789	623
780	275	157	800	410
800	567	75	536	37
609	765	399	326	59
554	735	400	201	685
343	525	401	99	483
579	470	321	198	245
6	246	217	708	373

Class Intervals	Frequency	Relative Frequency	Cumulative Frequency	Cumulative %

    Total scores

7. Create a histogram of the frequencies from the table above:

8. Describe the shape of the histogram using terminology from the chapter.

Use the following information to answer questions 9–12:

A student in an introductory physiology class thought there was a pattern between the temperature and the number of minutes she spent outside. To determine if this was more than her imagination, she decided to keep track of that information over the course of 2 weeks. Her data are presented in the following table:

	Temperature	Number of minutes spent outside
Day 1	65	21
Day 2	67	25
Day 3	70	34
Day 4	65	26
Day 5	71	22
Day 6	70	80
Day 7	68	100
Day 8	69	30
Day 9	70	25
Day 10	68	36
Day 11	68	40
Day 12	69	50
Day 13	71	85
Day 14	70	92

9. Create a scatterplot of the data.

10. Identify the dependent and independent variables.

11. Interpret the scatterplot.

12. *Critical thinking:* What else might be accounting for the number of minutes spent outside?

Use the following information to answer questions 13–16:

A sociology professor wants to show his students that studying can improve their grade. To do so, he creates a graph of hours studied and grade received on the first exam.

Hours	Grade	Hours	Grade
10	99	7	87
5	78	6	80
6	80	4	80
2	78	5	78
0	69	2	70
9	90	5	94
8	90	4	92
10	97	5	97
3	80	5	85
4	84	3	78

13. What is the most appropriate way to graph these data?

14. Graph the data.

15. Identify the independent and dependent variables.

16. Interpret the graph. Does the teacher have evidence supporting his point?

Chapter Three Study Guide Answers

1. d

2. b

3. a

4. Would likely be negatively skewed with elderly people taking more medications daily. Kurtosis may also be evident, as folks may take medications in later years, but not in latest years.

5. a. bar
   b. histogram
   c. histogram
   d. scatterplot

6.

Class Intervals	Frequency	Relative Frequency	Cumulative Frequency	Cumulative %
701–800	7	.175	40	100.00
601–700	3	.075	33	82.50
501–600	5	.125	30	75.00
401–500	5	.125	25	62.5
301–400	6	.15	20	50.00
201–300	5	.125	14	35.00
101–200	3	.075	9	22.50
1–100	6	.15	6	15.00
Total scores	40			

7.

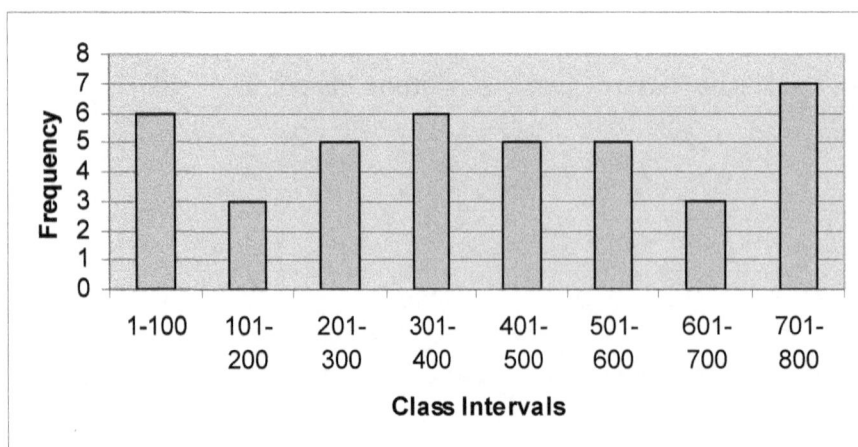

8. The graph is asymmetrical, with extreme scores at both the positive and negative end skewing it in both directions. It also appears slightly kurtotic, with a flattening at the center of the distribution.

9.

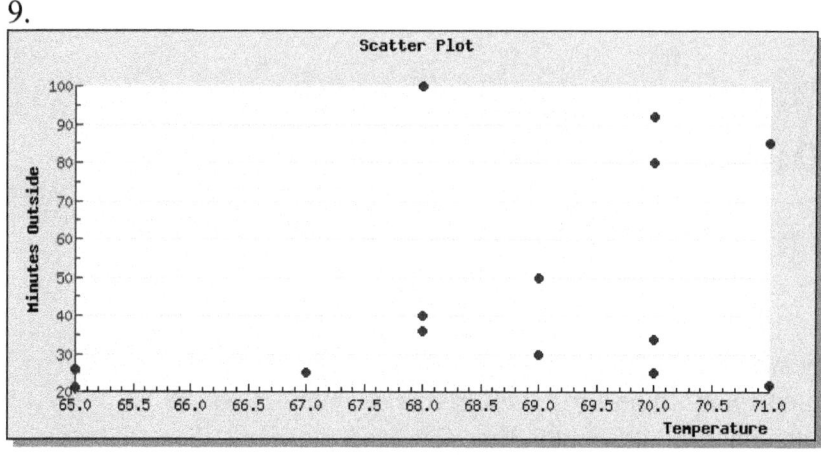

10. Independent variable—temperature; Dependent variable—amount of time spent outside

11. The scatterplot shows a weak positive correlation between temperature and minutes spent outside, suggesting that as the temperature increases, the minutes spent outside also increase.

12. The amount of time spent outside may be impacted by the day of the week (greater number of minutes occur on days 6 and 7, which may coincide with the weekend).

13. A scatterplot would be most appropriate.

14.

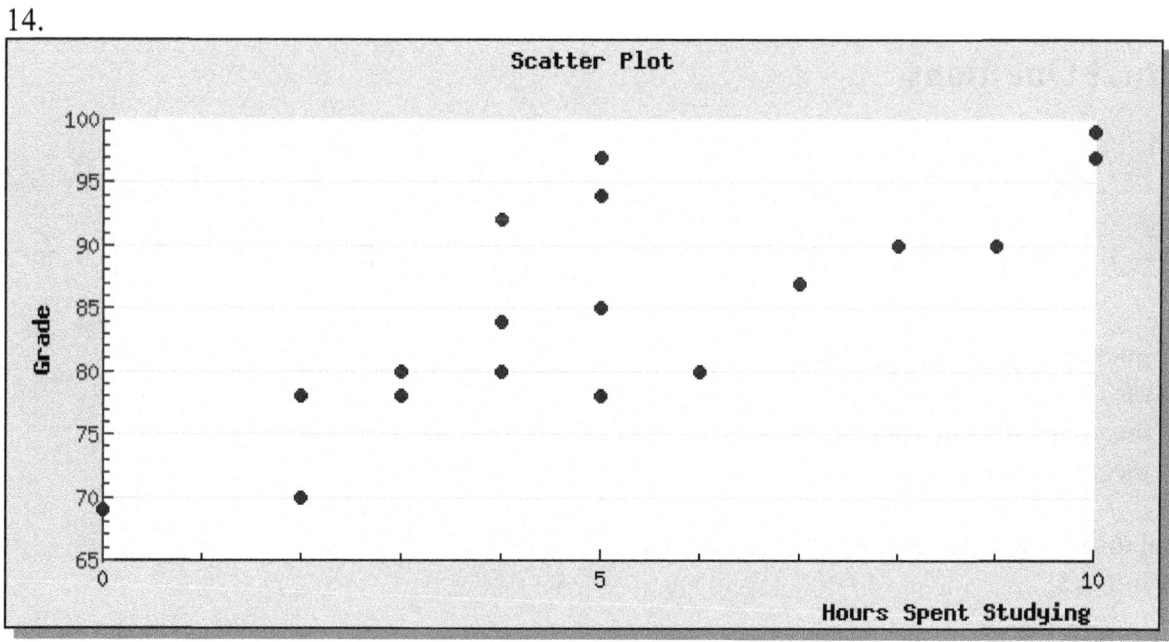

15. Independent variable—time spent studying; Dependent variable—grades

16. The graph shows a moderate positive correlation, suggesting that the number of hours spent studying is positively correlated with grade received. The teacher has data supporting his point.

Chapter Four

Study Guide Questions

Notes to Students

The math in this chapter is fairly simple, but you'll have to apply the notation concepts that you learned in the previous chapter correctly to perform well in this chapter. If you're finding you are having difficulty with this material, it is likely that you need to review the notation information or you need to become more familiar with your calculator. If you are struggling with your calculator, you should consider purchasing an inexpensive and simple one that has the symbols we recommended in Chapter 1 of the textbook. Simple calculators are easier to use and will increase the chances that you will use it correctly and get the right answer! This is worth spending an extra $6–8 USD if your current calculator is cumbersome or lacks the necessary buttons. Having the statistical symbols built in to it really saves a lot of time! If calculators are allowed on exam day, be sure that you have used it enough to be comfortable with it for the test. That usually means you need to learn how to store numbers in it to fully use the memory function. Again, your instructor may be able to help you with the practical aspects of using your calculator and fellow students are often willing to lend a hand as well.

Conceptual Questions

Multiple Choice Questions

1. $\mu = \dfrac{\sum X_i}{N}$ denotes the _____ of the _____ while $\overline{X} = \dfrac{\sum X_i}{n}$ denotes the mean of the _____.
  - (a) median, population, population
  - (b) mean, sample, population
  - (c) mean, population, sample
  - (d) mean, sample, population

2. A normal distribution curve would be described as _____.
  - (a) bimodal
  - (b) unimodal
  - (c) skewed
  - (d) multimodal

Short Answer Questions

For the following distributions, would you use the mean or the median to represent the central tendency? Why?

3. 2,3,8,5,7,3

4. 10,12,15,13,19,22

5. 1.2, 0.8, 1.1, 0.6, 25

6. Why are descriptive statistics important?

7. How does variance differ from standard deviation? Provide both an explanation and the appropriate notation.

8. $\sum \left(X_i - \overline{X} \right)$ represents _____ and is equal to _____.

9. Explain what SS means in your own words.

Calculation Questions

Use the following information to answer questions 10–13:

A psychologist interested in the dating habits of college undergraduates samples 10 students and determines the number of dates they have had in the last month. These are the data:

1	8	12	3	8
14	4	5	8	16

10. Calculate the mean, median, mode, and range.

11. Calculate the SS.

12. Estimate the standard deviation of the population.

13. Estimate the variance of the population.

Use the following information to answer questions 14–17:

A clinical psychologist is interested in the average number of sessions attended by a sample of patients seen at the on-campus training clinic. The following data are obtained:

10	7	14	14
25	15	23	20
31	40	10	25
15	62	8	14
39	37	51	12

14. Calculate the mean, median, mode, and range.

15. Calculate the SS.

16. Estimate the standard deviation of the sample.

17. Estimate the variance of the sample.

Use the following information to answer questions 18–21:

A public health professor gives a quiz to her class and records the following scores:

13	11	11	9
12	13	16	14
11	10	8	13
20			

18. Calculate the mean, median, mode, and range.

19. Calculate the Sum of Squares (SS).

20. Calculate the standard deviation (assume that your population consists of only the students in this public health class).

21. Calculate the variance.

Chapter Four Study Guide Answers

1. c

2. b

3. The mean, because there are no extreme scores

4. The mean, same as (a)

5. The median, because the distribution contains an extreme score (25)

6. They can help determine if inferential statistics are appropriate, and provide summaries of raw data

7. **Variance:** the squared average deviation of a score from the mean. $\sigma^2 = \dfrac{\sum X^2 - \dfrac{(\sum X)^2}{N}}{N}$

Standard deviation: the average deviation of a score from the mean. $\sigma = \sqrt{\dfrac{\sum X^2 - \dfrac{(\sum X)^2}{N}}{N}}$.

8. The sum of deviations around the mean; zero

9. The sum of the squared deviations around the mean; the sum of the squared distance that each value is from the mean

10. mean = 7.90, median = 8, mode = 8, range = 15

11. SS = 214.9

12. s_{n-1} = 4.88648 = 4.89

13. s^2 = 23.877778 = 23.88

14. mean = 23.6, median = 17.5, mode = 14, range = 55

15. 12.38

16. s = 3.00098 = 3.00

17. s^2 = 9.0059 = 9.01

18. mean = 12.38, median = 12, modes = 11 and 13, range = 12

19. SS = 117.077

20. $s = 3.123$

21. $s^2 = 9.756$

Chapter Five

Study Guide Questions

Notes to Students

This chapter is a good example of how much easier the math is compared to the concepts in statistics. You probably won't have trouble with the math if you work enough problems for practice, but understanding that 100% of the probability lies below the curve is a bit more difficult. Make sure that you reread this material until you understand it because it will be critical for understanding later chapters. This is a place where you can do yourself a lot of good by not missing class when these concepts are covered, and by reading your textbook carefully. Don't be afraid to ask questions in class or to talk this over with other students until you can work the z-score problems forward (finding a z-score value) and backward (solving for X), *and* you can interpret how the area under the curve relates to probability.

Conceptual Questions

Multiple Choice Questions

1. In a normal curve, the inflection points occur at _____.
>   (a) $\mu \pm 1\sigma$
>   (b) $\pm 1\sigma$
>   (c) $\mu \pm 2\sigma$
>   (d) μ

Short Answer Questions

Define the following terms:

2. Asymptotic:

3. z-score:

4. "Probability": (Note: use words rather than symbols.)

5. "Probability": (Note: use an equation rather than words.)

6. Why are standardized scores important? What information do they relay?

7. Describe the differences between theoretical and empirical determination of the probability of an event.

Calculation Questions

8. Given the follow set of normally distributed sample data:

| 10 | 12 | 16 |
| 18 | 19 | 21 |

Convert each raw score to a z-score.

Use the following data for questions 9 and 10:

A testing bureau reports that the mean for the population of Graduate Record Exam (GRE) scores is 500 with a standard deviation of 90. The scores are normally distributed.

9. What percentage of scores falls below 667?

10. What percentage of scores lies between 460 and 600?

Use the following data for questions 11–13:

A public health researcher has kept a monthly record of the expense of cigarettes in a large smoking population and finds that the amount of money spent on cigarettes per month forms a normal distribution with a mean of $60.00 and a standard deviation of $4.30.

11. The raw score corresponding to a z-score of 0.00 is _____.

12. The raw score corresponding to a z-score of -1.51 is _____.

13. The raw score corresponding to a z-score of 2.02 is _____.

Use the following data for questions 14–16:

An obesity researcher is using a mouse model to look at the impact of diet soda consumption on weight. Her data suggest that, on average, the mice will consume 85oz of soda per week, with a standard deviation of 7.5oz.

14.   The raw score corresponding to a z-score of -.9 is _____.

15.   What percentage of the population consume between 75 and 90 ounces per week?

16.   How many ounces per week do mice at the 95th percentile consume?

17. Assume you are rolling two fair dice once. What is the probability of obtaining a sum of 5?

18. Assume you are rolling two fair dice once. What is the probability of obtaining a 1 and a 6 (order doesn't matter)?

Use the following data for questions 19 and 20:

Suppose a researcher has three different experimental tasks for students to do as part of a study on attention and executive function: the Stroop task, the Wisconsin Card Sort (WCST), and a Letter-Number sequencing task. Participants are randomly assigned with regards to what order they will complete the tasks.

19. What is the probability of order being: WCST, Stroop, Letter-Number?

20. What is the probability notation for question 18?

Chapter Five Study Guide Answers

1. a

2. Asymptotic = approaching a probability of zero as z increases or decreases to infinity: this term captures the concept that a probability can never fall completely to zero; there is *always* some finite probability of an even occurring, however unlikely.

3. z-score = a z-score is an actual score that has been converted to an artificial ("standardized") score that reflects the original scores relative position in the distribution of all scores in the sample.

4. Probability = a probability is the number of theoretical or actual occurrences of an event as a function of the total number of possible events, or outcomes. It is the "likelihood" of a certain event occurring.

5. $p(A)$ = # of A events / # total occurrences

6. Standardized scores are important because they show the relative status of a score in the distribution. It helps you understand the meaning of a score by comparing it to other scores in a more broadly interpretable way.

7. A theoretical probability is a probability calculated on the basis of a theoretical knowledge of the number of possible outcomes for an event, while an empirical probability is based on observing the actual number of occurrences of an event or outcome.

8. $10 = -1.41$
       $12 = -0.94$
       $16 = 0.00$
       $18 = 0.47$
       $19 = 0.71$
       $21 = 1.18$

9. z-score for 667 is 1.85. From col C = 3.22, so 100 - 3.22 = 96.78%

10. 460 → z-score = -0.44 → col B = 17.00
       600 → Z-score = 1.11 → col B = 36.65
       17.00 + 36.65 = 53.65%

11. X = 60.00 + 4.30(0) = 60.00

12. X = 60.00 + 4.30(-1.51) = 53.507 = 53.51

13. X = 60.00 + 4.30(2.02) = 68.686 = 68.69

14. X = 85 + 7.5(-.9) = 78.25

15. 75 → z-score = -1.333 → Column B → 40.82
       90 → z-score = .666 → Column B → 24.54
       40.82 + 24.54 = 65.36%

16. X = 85 + 7.5 (1.645) = 97.34 oz

17.
p(rolling a 2 on one die and a 3 on the other) = 1/6 x 1/6 = 0.02777777778
p(rolling a 3 on one die and a 2 on the other) = 1/6 x 1/6 = 0.02777777778
p(rolling a 1 on one die and a 4 on the other) = 1/6 x 1/6 = 0.02777777778
p(rolling a 4 on one die and a 1 on the other) = 1/6 x 1/6 = <u>0.02777777778</u>
$$\qquad\qquad\qquad\qquad\qquad\qquad 0.11111111111$$

Therefore, the probability of obtaining a sum of 5 on one roll of two fair dice is 11.11%.

18.
p(rolling 1 on one die and 6 on the other) = 1/6 x 1/6 = 0.02777777778
p(rolling 6 on one die and 1 on the other) = 1/6 x 1/6 = <u>0.02777777778</u>
$$\qquad\qquad\qquad\qquad\qquad\qquad 0.05555555556$$

Therefore, the probability of rolling a 1 and a 6 on one roll of two fair dice is 5.56%.

19. p(WCST) = .33333
$\quad p$(Stroop) = <u>.55555</u>
$$\qquad\qquad .16667$$

Therefore, the probability of order being WCST, Stroop, Letter-Number is .166667.

20. p(WCST and Stroop and Letter-Number) = p(WCST) * p(Stroop|WCST) * p(Letter-Number|WCST+Stroop)

Chapter Six

Study Guide Questions

Notes to Students

In this chapter you calculate your first inferential tests. Again, the math will be easy because it is simply an extension of the z-score formula that you used in Chapter 5. The new concept here will be sampling distributions and how we can use them to evaluate our statistical results. This is another important concept because you'll use sampling distributions throughout the rest of the course.

If you understood Chapters 1–5 thoroughly, then this chapter will merely be an extension of what you have already learned. If you feel the least bit lost, this is a good time to seek outside help from your instructor, a teaching assistant, a statistical tutor, or a study group. Don't feel intimidated to ask for help and do not assume that you are the only person in the class that might be feeling the least bit lost! Statistics and probability are a novel way of looking at the world and most students need a bit of time before it becomes natural. In fact, it's not unusual for the "light bulb" to go on for our students near the end of the course. Once it does, they usually realize that they were missing some key early concept that has now been clarified or that they were making it all harder than necessary. One of the best ways to learn this material is to teach it to others, so study groups of serious students can be a big help. Our best students have volunteered as undergraduate peer tutors in subsequent courses and found that they learned even more the second time they sat through the course and explained the concepts to others. This is a key reason why we feel that you cannot review too much in a statistics course! There are nuances to the conceptual material in statistics that you might miss if you only read the chapter once.

Conceptual Questions

True/False Questions

1. The sampling distribution of the mean is always normally distributed.

2. The sampling distribution of the mean varies with the size of the sample.

3. To use the z test, μ and σ must be defined.

4. As N increases, the variability of the sampling distribution of the mean increases.

5. To determine whether a particular sample came from the population of interest, you:
   (a) Calculate the appropriate statistic
   (b) Evaluate the statistic based on the appropriate sampling distribution
   (c) None of the above
   (d) a & b

Short Answer Questions

6. If we're a provided a sample of $N = 15$, we can't outright assume that statistics can be computed. Why?

7. Explain what each piece of the formula $z = \dfrac{x - \mu}{\sigma}$ represents.

Calculation Questions

Use the following data for questions 8–11:

Suppose we know that the population of undergraduate psychology students at UW have a mean IQ of 110, with a standard deviation of 5. Assume these data are normally distributed. We draw the following scores from a 315 stats class as a sample of our population:

118	119	110	110	112
115	112	112	115	110

8. Calculate $\overline{X}$ for this sample.

9. Calculate $\mu_{\overline{x}}$.

10. Calculate $\sigma_{\overline{x}}$.

11. What is the probability that our sample comes from the original population? Briefly describe.

Use the following data for questions 12–14.

A golfer is interested in finding out where his score lies in the normal distribution of golf scores at a tournament. The mean is 71.44, the standard deviation is 15.94, and the scores are normally distributed.

   12. If the golfer scored a score of 65, what percentage of the golfers scored *below*?

   13. Draw the *z*-distribution, shading the portion of the distribution representing golfers that scored below 65.

   14. Local pros wanted to see how many golfers' scores were in the top 10% of those at the tournament. What would be the cut-off score for qualifying for this percentile?

Combined (Concepts and Calculation) Questions

Use the following data for questions 15–17:

A professor has been teaching statistics for many years. His records show that the overall mean for the final exam scores is 82 with a standard deviation of 10. The professor believes that this year's class is superior to his previous ones. The mean for the final exam scores for this year's class of 65 students is 87.

   15. Calculate σ_{x}.

   16. Calculate $z_{obt.}$

   17. Interpret $z_{obt.}$ What can you conclude?

Use the following data for questions 18–20:

On the basis of her newly developed technique, a student believes she can reduce the amount of time people with post-traumatic stress disorder (PTSD) spend in therapy. As director of clinical training at a local college, you agree to let her try her method on 20 randomly selected patients diagnosed with PTSD. The mean duration that PTSD patients spend in therapy through your college is 85 weeks with a standard deviation of 15 weeks. The scores are normally distributed. The results of the experiment show that the patients treated by the student are in therapy for a mean duration of 78 weeks with a standard deviation of 20 weeks.

18. Calculate $\sigma_{\bar{x}}$.

19. Calculate z_{obt}.

20. What can you conclude about the student's technique?

Chapter Six Study Guide Answers

1. False

2. True

3. True

4. False

5. d

6. The sample size is too small to invoke CLT—would need larger N. Or need to be told that raw scores are normally distributed.

7. X represents a single score, μ represents the population mean, and σ is the standard deviation of the population.

8. 113.3

9. Would be the same—110

10. $5/\sqrt{10} = 1.581$

11. $z_{\bar{x}} = (113.3-110) / 1.581 = 2.09$; Column C = .0183; highly unlikely that this sample is drawn from our population—less than 1.8%

12. $z = \dfrac{65 - 71.44}{15.94} = \dfrac{-6.44}{15.94} = -.40401$

Table A: .1554 = 15.54% of players scored below a 65 during the NEC.

13.

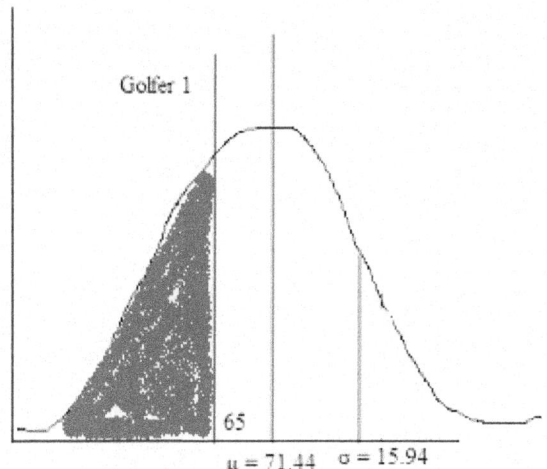

14. $X = 71.44 + 15.94\,(2.33) = 108.58$

15. $z_{obt} = \dfrac{\text{X}_{mean} - \mu}{\sigma_x}$ $\qquad \sigma_x = 10/(65)^{1/2} = 1.24035$

16. $z_{obt} = \dfrac{87 - 82}{1.24035} = 4.03113 = \mathbf{4.03}$

17. Based on the table, this is an unlikely result due to chance. It appears that this year's class has significantly higher scores than those of previous years' classes.

18.
$z_{obt} = \dfrac{\text{X}_{mean} - \mu}{\sigma_x}$ $\qquad \sigma_x = 15/(20)^{1/2} = 3.354102$

19. $z_{obt} = \dfrac{78 - 85}{3.354102} = \mathbf{-2.09}$

20. Based on the table, this is an unlikely result due to chance. There appears to be a difference in the duration of time spent in therapy between PTSD patients treated through the college and those treated using the new technique.

Chapter Seven

Study Guide Questions

Notes to Students

This chapter is completely conceptual. There is no math in this chapter, but there are some important concepts that you'll be using heavily from now on. This is another point in the course where attending class and rereading the textbook will help you grasp the material in this chapter, as well as the material that you'll be discussing for the rest of the term. We would recommend that you do all of the exercises in the textbook as well as the exercises in the study guide for this chapter. Most students won't understand the material fully unless they work enough problems. If you need more problems to work prior to taking an exam, remember that you can consult other textbooks in the library. You could also look at a few websites. We recommend the following website: http://davidmlane.com/hyperstat/, which contains a great deal of information on statistics at the undergraduate level. If you cannot find this link or you wish to find another link, be sure to use a reliable site from a university or college professor teaching statistics. This is not a guarantee that there will not be any typographical errors, but the information is highly likely to be reliable.

Conceptual Questions

True/False Questions

1. The critical region for rejection of H_0 is the area under the curve that contains all the values of the statistic that fail to allow rejection of H_0.

2. The critical value of a statistic is the value of the statistic that bounds the critical region for rejection of H_0.

3. Power is the sensitivity of the experiment to detect a real effect of the independent variable, if there is one.

4. If H_0 is false, a high level of power increases the probability we will reject it.

5. Increasing N increases the real effect of the independent variable.

6. Alpha and Beta are inversely related.

Short Answer Questions

Use the following information to answer questions 7–10:

A researcher wants to investigate the effects of caffeine on cognitive ability in students. She randomly selects 15 undergraduate students to participate in the study and assigns 7 to the control condition (no caffeine) and 8 to the experimental condition (2 cups of coffee). She waits 20 minutes and then administers a test to each group of students. The mean test score for the control condition is 8.4 ($s = 0.79$), while the mean test score for the experimental condition is 5.3 ($s = 0.51$). She sets her *p*-value to .05.

7. What is the null hypothesis?

8. What is the alternative hypothesis?

9. Is the researcher going to perform a one-tailed or two-tailed test?

10. Suppose the researcher runs the experiment and gets a *p* value of .042. What does this *p*-value mean, and how should she interpret her null hypothesis?

11. What factors affect power (list all 6)?

12. Describe what it means to set your α .05.

13. Complete the following table:

Statistical Decision	Null Hypothesis True (No Real Effect)	Null Hypothesis False (Real Effect)
Fail to reject the null hypothesis		
Reject null hypothesis		

14. When would you use a non-parametric statistical test? Are these tests more or less powerful than parametric tests?

15. Many researchers use sampling without replacement to select subjects, which violates an assumption for inferential statistics. What allows researchers to continue with this practice?

Use the following information to answer questions 16–20:

Dr. Shen is studying the effect of ginseng on activity in rats. She believes that ginseng will increase activity, and she wants to test this hypothesis. Activity levels (as measured in a running wheel) in rats are known to be normally distributed and to have a population mean of 36 rpm and a population standard deviation of 9.

16. Is the alternative hypothesis for this study directional or non-directional?

17. State the researcher's null hypothesis (in words).

18. State the researcher's alternative hypothesis (in words).

19. If you calculate a p value of .057, did ginseng have an impact on the activity level in rats? Use an alpha level of .05.

20. What error might you be making based on your conclusion in question 18?

Chapter Seven Study Guide Answers

1. False

2. True

3. True

4. True

5. False

6. True

7. H_O: There is no effect of caffeine on cognitive ability.

8. H_1: Caffeine affects cognitive ability.

9. Two-tailed test

10. This *p*-value is less than her α of .05, meaning she can reject the null hypothesis and assume that caffeine does have an effect cognitive ability.

11. Magnitude of the real effect of your independent variable, sample size, alpha, experimental design, statistical test, and type of hypothesis

12. Setting an α to .05 means the alternative hypothesis to a research question would only occur in 5% or fewer cases under normal circumstances, meaning the likelihood that it occurred due to chance is low. Therefore, it is more likely that the variables in an experiment affected the outcome.

13.

Statistical Decision	Null Hypothesis True (No Real Effect)	Null Hypothesis False (Real Effect)
Fail to reject the null hypothesis	**Correct Decision**	**Type II error**
Reject null hypothesis	**Type I error**	**Correct Decision**

14. Non-parametric statistical tests are used when the assumptions about the underlying parameters of the population have been violated. Non-parametric tests are less powerful than parametric tests.

15. Researchers can use sampling without replacement because if the sample is small relative to the total population, the effect on the assumptions underlying the statistical test(s) is minor.

16. Directional, since the researcher is hypothesizing that ginseng will increase activity

17. H_O: There is no effect of ginseng on activity level in rats.

18. H_1: Ginseng will increase activity level in rats.

19. Since p of .057 is > .05, it does not appear that ginseng has an impact on activity level in this sample of rats.

20. This may be a Type II error, in that we may have failed to reject the null hypothesis when there is actually a real effect of ginseng.

Chapter Eight: Single Sample Tests

Study Guide Questions

Notes to Students

In this chapter you'll also be making heavy use of the concepts from Chapters 5–7. In particular, hypothesis testing and critical values will be important concepts that will be built upon in this chapter. Lastly, you'll have your first experience at choosing which statistical test is appropriate given two similar tests (Single sample z-test and the Single sample t-test).

Conceptual Questions

True/False Questions

1. For a given N, the interval that bounds the population mean at the 99% confidence level is greater than that which bounds the population mean at the 95% confidence level.

2. The t test differs from the z test in that with the t test we estimate σ.

Short Answer Questions

3. What factor impacts the t-distribution?

4. Describe effect size. Why is this a useful number to calculate?

Calculation Questions

5. (a) A sample set of 29 scores has a mean of 76 and a standard deviation of 7. Can we accept the hypothesis that the sample is a random sample from a population with a mean greater than 72? Use $\alpha = .01_{1tailed}$ in making your decision.

(b) Calculate the 95% confidence interval for the population mean.

6. A professor has been teaching statistics for many years. His records show that the overall mean for the final exam scores is 82 with a standard deviation of 10. The professor believes that this year's class is superior to his previous ones. The mean for the final exam scores for this year's class of 65 students is 87. What do you conclude? Use $.05_{1tailed}$.

Combined (Concepts and Calculation) Questions

Use the following information to answer questions 7–11:

On the basis of her newly developed technique, a student believes she can reduce the amount of time people with post-traumatic stress disorder (PTSD) spend in therapy. As director of clinical training at a local college, you agree to let her try her method on 20 randomly selected patients diagnosed with PTSD. The mean duration that PTSD patients spend in therapy through your college is 85 weeks with a standard deviation of 15 weeks. The scores are normally distributed. The results of the experiment show that the patients treated by the student are in therapy for a mean duration of 78 weeks with a standard deviation of 20 weeks.

7. What is the alternative hypothesis? (assume a non-directional hypothesis)

8. What is the null hypothesis?

9. What can you conclude about the student's technique? Use an alpha level = $.05_{2tailed}$.

10. What error might you be making by your conclusion in part (c)?

11.
      (a) If you changed the alpha level to .08, how would this affect the power of the experiment?
      (b) How would this affect your chances of making a Type II error?

Use the following information to answer questions 12–16:

A well-known psychology graduate program claims that their PhD graduates get higher-paying jobs than the national average. Last year's figures for salaries paid to all graduates with a psych PhD on their first job showed a mean of $6.20 per hour. A random sample of 10 graduates from last year's class of psychology PhDs showed the following hourly salaries for their first job:

$5.40	$6.30	$7.20	$6.80	$6.40
$5.70	$5.80	$6.60	$6.70	$6.90

12. What is the alternative hypothesis? (assume a non-directional hypothesis)

13. What is the null hypothesis?

14. What can you conclude about the psychology graduate program's claim? Use an alpha level = $.05_{2tailed}$.

15. What error might you be making by your conclusion in part (c)?

16. If you were only concerned with evaluating whether salaries *increased*, how would this affect the power of your experiment?

Use the following information to answer questions 17–20:

You wanted to estimate the mean number of vehicles crossing a busy bridge in your neighborhood each morning during rush hour for the past year (you wacky thing, you!). To accomplish this, you stationed yourself and a few assistants at one end of the bridge on 18 randomly selected mornings during the year and counted the number of vehicles crossing the bridge in a 10-minute period during rush hour. You found the mean to be 125 vehicles per minute with a standard deviation of 32.

17. Calculate the 95% confidence interval for the population mean.

18. State your conclusion from part (a) in words.

19. Calculate the 99% confidence interval for the population mean.

20. What do you notice about the difference between your answers for (a) and (c)? How do you explain this difference?

Chapter Eight Study Guide Answers

1. False

2. True

3. The t-distribution is impacted by the degrees of freedom, which is related to the sample size.

4. Effect size is a standardized measure of difference between two (or more) group means. Because effect size is a standardized value, it allows for comparison across studies.

5. (a)

$N = 29$ $\mu = 72$ $x_{mean} = 76, s = 7$

$t_{obt} = \dfrac{x_{mean} - \mu}{s_x}$ $s_x = 7/(29)^{1/2} = 1.2998673$

$t_{obt} = \dfrac{76 - 72}{1.2998673} = 3.077237 = \mathbf{3.08}$ $df = 28, \quad t_{crit} = |2.467|$

$|3.08| > |2.47|$; Reject H_0—The sample appears to be a random sample from a population with a mean greater than 72.

(b) $x_{mean} - s_x(t_{crit}) < \mu < x_{mean} + s_x(t_{crit})$
$= 76 - 1.2998673(2.0484) < \mu < 76 + 1.2998673(2.0484)$
$= 73.33735 < \mu < 78.66265$
$= \mathbf{73.34 < \mu < 78.66}$

6. $N = 65$ $\mu = 82, \sigma = 10$ $x_{mean} = 87$

$z_{obt} = \dfrac{x_{mean} - \mu}{\sigma_x}$ $\sigma_x = 10/(65)^{1/2} = 1.24035$

$z_{obt} = \dfrac{87 - 82}{1.24035} = 4.03113 = \mathbf{4.03}$ $z_{crit} = |1.645|$

$|4.03| > |1.645|$; Reject H_0—This year's class has significantly higher scores than those of previous years' classes.

7. H_1: There is difference in the duration of time spent in therapy between PTSD patients treated through your college and those treated using the new technique.

8. H_0: There is no difference in the duration of time spent in therapy between PTSD patients treated through your college and those treated using the new technique.

9. $N = 20$ $\qquad$ $\mu = 85$, $\sigma = 15$ $\qquad$ $x_{mean} = 78$, $s = 20$

$z_{obt} = \dfrac{x_{mean} - \mu}{\sigma_x}$ $\qquad$ $\sigma_x = 15/(20)^{1/2} = 3.354102$

$z_{obt} = \dfrac{78 - 85}{3.354102} = \textbf{-2.09}$ $\qquad$ $z_{crit} = |1.96|$

$|2.09| > |1.96|$; Reject H_0—There appears to be a difference in the duration of time spent in therapy between PTSD patients treated through the college and those treated using the new technique.

10. Type I error (rejecting the null hypothesis when it is actually true)

11.
  (i) Increasing your alpha level increases power, and
  (ii) reduces your chances of making a Type II error.

12. H_1: There is a difference between the salaries of this year's class of psychology PhDs and the national average.

13. H_0: There is no difference between the salaries of this year's class of psychology PhDs and the national average.

14. $N = 10$ $\qquad$ $\mu = 6.2$ $\qquad$ $x_{mean} = 6.38$, $s = .581$

$t_{obt} = \dfrac{x_{mean} - \mu}{s_x}$ $\qquad$ $s_x = .581/(10)^{1/2} = 0.183728$

$t_{obt} = \dfrac{6.38 - 6.2}{0.183728} = 0.97971 = \textbf{0.98}$ $\qquad$ $t_{crit} = |2.262|$

  $|0.98| < |2.262|$; Accept H_0—There is no difference between the salaries of this year's class of psychology PhDs and the national average.

15. Type II error (accepting the null hypothesis when it is actually false)

16. This would make your experiment directional (1-tailed), and would thus increase the power.

17. $x_{mean} - s_x(t_{crit}) < \mu < x_{mean} + s_x(t_{crit})$
  $= 125 - [32/(18)^{1/2}](2.1098) < \mu < 125 + [32/(18)^{1/2}](2.1098)$
  $= 109.08689 < \mu < 140.91311$ $\quad = \textbf{109.09} < \mu < \textbf{140.91}$

18. There is a 95% probability that the interval from 109.09 to 140.91 contains the population mean.

19. $x_{mean} - s_x(t_{crit}) < \mu < x_{mean} + s_x(t_{crit})$

$= 125 - [32/(18)^{1/2}](2.8982) < \mu < 125 + [32/(18)^{1/2}](2.8982)$

$= 103.140407 < \mu < 146.859593$

$= \mathbf{103.14 < \mu < 146.86}$

20. The 99% confidence interval is larger than the 95% CI. Because we are only willing to accept being wrong about our interval containing the population mean only 1% of the time instead of 5%, we are going to have a larger interval; i.e., there is a wider range of numbers that fall within the 99% CI, which increases our probability of encompassing μ.

Chapter Nine: Two Sample Tests

Study Guide Questions

Notes to Students

On first glance the formulas in this chapter may look a bit scary, but if you look carefully you'll see that it's all basic math (addition, division, square root, etc.). The reason it looks scary is that there are multiple basic math steps in one problem. At this point it will be important to review your skills on doing the math in the correct order. Recall that operations in parentheses are done first and that you should not round until you reach the end of the problem. If you are using your calculator well, you can save the intermediate products or sums in your calculator's memory. If this is too confusing or you need to show your work for the assignment or exam, then write down the intermediate answers carefully. Sloppy work might result in misreading some of these intermediate numbers and an error in your final answer. Again, practice makes perfect with this math. Students who work sufficient problems come up with a system of working the problem that results in minimal errors and maximum speed. You may find that you prefer this more mathematical chapter over the conceptual chapters! This will be particularly true if you've done the practice to excel in this chapter.

The other key skill to practice in this chapter is in determining which of the two sample tests to perform (independent t-test or paired t-test). Here you must combine what you know about experimental design (between-groups designs versus within-groups designs) with the statistics you have learned, and then you'll be able to choose the appropriate test to calculate. We'll also ask you to distinguish between these two new statistical tests and the two you learned in the previous chapter by giving you story problems where any of the four tests could be the answer. This is the most important skill to learn if you are ever going to use or evaluate statistics in the future. What is the appropriate test given the experimental design and can you determine that from a paragraph describing the research study? If you can, then you can rely on computers for the actual calculations!

Conceptual Questions

Multiple Choice Questions

1. Student's t-test for correlated groups really reduces to _____.
    (a) student's t-test for single samples using difference scores
    (b) student's t-test for independent groups
    (c) none of the above

2. As compared to a dependant samples t-test, independent samples t-tests are

_____.

    (a) more powerful
    (b) less powerful
    (c) equally as powerful
    (d) none of the above

Short Answer Questions

3. What is the homogeneity of variance assumption? When can you assume HOV?

4. What are two major drawbacks to using a within-groups design?

Combined (Concepts and Calculation) Questions

Use the following information to answer questions 5–7:

A political candidate wishes to determine if endorsing increased social spending is likely to decrease her popularity. She has access to data on the popularity of several other candidates who have endorsed increased spending. The data were available both before and after the candidates announced their positions on the issue. The data are as follows:

Popularity Ratings

Candidate	Before	After
1	42	43
2	41	45
3	50	56
4	52	54
5	58	65
6	32	29
7	39	46
8	42	48
9	48	47
10	47	53

5. State the hypotheses.

6. What is the value of t_{obt}?

7. What might the candidate conclude using $\alpha = .01$ (1tail)?

Use the following information to answer questions 8 and 9:

A biologist believes that temperature affects the croaking (noise making, not dying) behavior of frogs. Laboratory frogs are randomly divided into 2 groups and placed in identical terrariums. The control group of frogs is kept at a constant 22°C. The experimental group is kept at a temperature of 30°C. The number of frog croaks emitted over a 10-minute period are counted. The data are indicated below:

Number of Croaks

22°	30°
23	30
30	32
31	36
28	39
26	30
12	18
19	25
18	26

8. State the hypotheses.

9. What is the value of t_{obt}?

Use the following information to answer questions 10 and 11:

A researcher is interested in studying the effects of alcohol on learning ability in rats. He randomly assigns 10 rats to an "alcohol group" and another 10 to a control group. The rats in the alcohol group receive 1 oz. of alcohol prior to being tested. Then all of the rats run through a maze and the number of their errors are recorded. Assume the data below are normally distributed.

Control	Alcohol Group
3	5
2	3
4	7
1	2
2	4
3	5
3	4
1	3
5	2
3	3

10. What is the appropriate statistical test and why?

11. Perform that test and state your conclusion on the null hypothesis.

12. A nurse is interested in determining whether immediate memory capacity is affected by sleep loss. Immediate memory is defined as the amount of material that can be remembered shortly after it has been presented. Twelve students are randomly selected and randomly assigned to two groups of 6 each. One of the groups is sleep-deprived for 24 hours before the material is presented. All subjects in the other group receive the normal amount of sleep (7–8 hours). The material consists of a series of slides, with each slide containing nine numbers. Each slide is presented for a short time interval (50 milliseconds), after which the subject must recall as many numbers as possible. The following are the results. The scores represent the percentage correctly recalled.

Normal Sleep	Sleep-deprived
68	42
73	62
72	68
65	73
70	57
73	49

What is the **most** appropriate statistical test? If assumptions are not met, please explain.

Use the following information to answer questions 13–15:

In a study of jury behavior, two samples of subjects were provided details about a trial in which the defendant was obviously guilty. Although group 2 received the same details as group 1, the second group was also told that some of the evidence had been withheld from the jury by the judge. Later the subjects were asked to recommend a jail sentence. The length of term suggested by each is presented here. Is there a significant difference between the two groups in their responses?

Group 1 scores: 4 4 3 2 5 1 1 4
Group 2 scores: 3 7 8 5 4 7 6 8

13. State the hypothesis, and select an alpha level.

14. What is your alpha level, number of tails, and degrees of freedom?

15. Compute the test statistic.

Answer questions 16–20 using the following story problem and data.

A researcher is interested in whether caffeine consumption affects problem-solving ability. She designs a study in which 10 subjects are presented with a puzzle to solve. The same 10 people then consume several ounces of caffeinated coffee and solve another puzzle. She records the time (sec) it takes for each subject to solve the puzzle in each condition:

Participant:	1	2	3	4	5	6	7	8	9	10
No caffeine:	140	118	303	137	151	198	148	187	211	132
Caffeine:	182	90	213	81	120	101	192	138	160	141

16. State the null hypothesis for this study.

17. State the alternative hypothesis for this study.

18. Is this a directional or non-directional hypothesis?

19. Is this research an independent groups or a repeated measures design? Why?

20. Which type of measurement scale do the data from this study represent (e.g., nominal, ordinal, interval, or ratio)?

Chapter Nine Study Guide Answers

1. a

2. b

3. HOV states that the variances of the underlying populations are equal or, in practical terms, not significantly different from one another, which is required to perform an independent samples t-tests. You can assume HOV if $\dfrac{\text{larger variance}}{\text{smaller variance}} \leq 4$.

4. Participants dropping out before data collection is complete, and experience effects.

5. H_0: Endorsing increased social spending has no effect on popularity ratings or increases popularity ratings.
H_1: Endorsing increased social spending decreasing popularity ratings.

6. Dbar = 3.5000 S_D = 3.5668 S_{Dbar} = 3.5668 / $\sqrt{10}$ = 1.1279

7. t_{obt} = 3.5 / 1.1279 = 3.103 t_{crit}(df = 9, α = .01) = 2.8214

Fail to Reject H_0: Endorsing increased social spending has no effect on popularity ratings or increases popularity ratings.

8. H_0: Temperature does not affect the croaking behavior of laboratory frogs.
H_1: Temperature affects the croaking behavior of laboratory frogs.

9. HOV   43. 9821/43.4286 = 1.01

S_w^2 = (7(43.9821) + 7(43.4286)) / (7+7) = 43.7054

t_{obt} = (23.3750 – 29.5000) / $\sqrt{43.7054\,(1/8 + 1/8)}$ = -1.8597

t_{crit} = 2.1448

Fail to Reject H_0: Temperature does not affect the croaking behavior of laboratory frogs.

10–11. **Independent t-test**

$\overline{X}_1$ = 2.7000 $\overline{X}_2$ = 3.8000
S_1 = 1.2517 S_2 = 1.5492
S_1^2 = 1.5667 S_2^2 = 2.4000

10. Independent t-test
Interval/ratio data, normal distribution, random selection, two unrelated samples, HOV

HOV = S_{larger}^2 / $S_{smaller}^2$ = 2.4000 / 1.5667 = 1.5319   Not violated

11. $S_w^2 = df_1(s_1^2) + df_2(s_2^2) / df_1 + df_2 = 9(1.2517^2) + 9(1.5492^2) / 9 + 9 = 1.9834$

$t_{obt} = \overline{X}_1 - \overline{X}_2 / \sqrt{(S_w^2 (1/n_1 + 1/n_2))}$

$\quad = 2.7000 - 3.8000 / \sqrt{1.9834(1/10 + 1/10)} = -1.7466$

$t_{crit}(df = 18, \alpha = .05, \text{2-tail}) = 2.101$

Fail to Reject H_0: Alcohol does not significantly affect maze learning in rats.

12. Independent t-test   Interval/ratio data, normal distribution, random selection, two unrelated samples, HOV

HOV $S^2_{larger} / S^2_{smaller} = 135.5000 / 10.1667 = 13.3278$ NOT Okay!

Since HOV is not satisfied you cannot continue with the independent t-test.

13. H_0: $\mu_1 - \mu_2 = 0$ For the population, knowing evidence has been withheld has no effect on the suggested sentence.

H_1: $\mu_1 - \mu_2 \neq 0$ For the population, knowledge of withheld evidence has an effect on the jury's response.

14. $\alpha = .05$ 2-tails $df = N-2 = 14$

15. $n_1 = 8$ $\overline{X}_1 = 3$ $s_1 = 1.5119$ $s_1^2 = 2.2857$ $S^2_w = 2.8571$

$n_2 = 8$ $\overline{X}_2 = 6$ $s_2 = 1.8516$ $s_2^2 = 3.4286$

$t_{obt} = -3.8451$

$t_{crit} (df = 14, \text{2-tails}, \alpha = .05) = 2.145$

Reject H_0: The knowledge of withheld evidence has an effect on the jury's response.

16. Caffeine consumption does not affect problem-solving ability or the results are due to chance.

17. Caffeine consumption affects problem-solving ability.

18. Non-directional

19. It is a repeated measures design because the study used the same participants for each condition. The participants served as their own controls.

20. Ratio or interval scale

Chapter Ten: ANOVA

Study Guide Questions

Notes to Students

In this chapter you will learn how to calculate an Analysis of Variance (ANOVA) test and that it is used when you have more than two groups or conditions in your study. This concept is essentially an extension of the t-tests that you have already learned about except that the ANOVA or F test is an overall test of significance. Because you have more than two groups, if you find a significant difference in the overall test, you still won't know which groups are really different from the others. Thus, we review planned and unplanned tests that are performed only if the ANOVA or F test is statistically significant. The math in the planned tests won't be problematic if you mastered the t-tests, and the math in the unplanned tests is similar. What most students struggle with is why we need to do the planned and unplanned tests and when we need to do them (only if the overall F test is significant). Reading the conceptual material carefully in this chapter should help you with this concept.

The other difficulty students may have at this point in the course is that they become overwhelmed with the number of different tests they have learned. This is a place where using the flowchart and/or reviewing your notes can help you keep all of these tests straight. It's best to do that now as the next chapter will include more tests! The most important thing to keep in mind is what test is used with a particular experimental design. Practice reading story problems to pick out how many groups or conditions there are in the experiment and whether or not the design is within-groups or between-groups. This will lead you to choose the appropriate test for any given situation.

Conceptual Questions

Multiple Choice Questions

1. Which of the following is (are) assumption(s) underlying the use of the F test?
a. The raw score populations are normally distributed.
b. The variances of the raw score populations are the same.
c. The mean of the population differ.
d. a and b
e. All of the above

2. Which of the following is (are) true about the F distribution?
a. It has no negative values.
b. It is positively skewed.
c. There are a family of F curves uniquely determined by df(numerator) and df(denominator).
d. The mean of the F distribution equals 0.
e. All of the above

f. a and b
g. a, b, and c

3. When analyzing data from experiments that involve more than two groups _____.

a. doing tests on all possible pairs of means decreases the probability of making Type 1 errors
b. doing t-tests on all possible pairs of means increases the probability of making Type 1 errors
c. it is generally permissible to do t-tests between all possible pairs of means and use t distribution
d. doing t-tests on all possible pairs of means increases the probability pf Type 2 errors

4. The alternative hypothesis evaluated by F_{obt} in the one-way ANOVA states that _____.

a. all conditions have the same effect
b. one or more of the conditions have different effects
c. $\mu_1 = \mu_2 = \mu_3 = \mu_k$
d. $\mu_1 \neq \mu_2 \neq \mu_3 \neq \mu_k$

5. Which of the following is (are) true?

a. df for $S_W^2 = N - k$
b. df for $S_B^2 = k - 1$
c. $df_T = N - 1$
d. All of the above

6. Which of the following would cause F_{obt} to increase?

a. An increase in the difference between the means
b. An increase in the within-groups variability
c. An increase in the magnitude of the independent variable's effect
d. a and b
e. a and c

7. S_B^2 is a measure of _____.

a. σ^2 alone
b. σ^2 + the effects of the independent variables
c. the variability between the means
d. b and c

Combined (Concepts and Calculation) Questions

Using the following information to answer questions 8–12:

Source	SS	df	s^2	F_{obt}	F_{crit}
Between	1253.68	3			
Within					
Total	5016.40	39			

8. Fill in the missing values.

9. How many groups are there in the experiment?

10. Assuming an equal number of subjects in each group, how many subjects are there in each group?

11. What is the value of F_{crit}, using $\alpha = .05$?

12. Is there a significant effect?

Use the following information to answer questions 13 and 14:

To test whether memory changes with age, a researcher conducts an experiment in which there are four groups of six subjects each. The groups differ according to the age of subjects. In group 1, the subjects are each 30 years old; group 2, 40 years old; group 3, 50 years old; and group 4, 60 years old. Assume that the subjects are all in good health and that the groups are matched on other important variables such as years of education, IQ, gender, motivation, and so on. Each subject is shown a series of nonsense syllables (a meaningless combination of three letters such as DAF or FUM) at a rate of one syllable every 4 seconds. The series is shown twice after which the subjects are asked to write down as many of the syllables as they can remember.

30 Years	40 Years	50 Years	60 Years
14	12	17	13
13	15	14	10
15	16	14	7
17	11	9	8
12	12	13	6
10	18	15	9

13. Use the analysis of variance with $\alpha = 0.05$ to determine whether age has an effect on memory.

Source	SS	df	s^2	F_{obt}	F_{crit}
Between					
Within			6.68		
Total	242.00				

14. If appropriate perform a multiple comparison test. The number of syllables remembered by each
     subject is shown here:

Use the following information to answer questions 15–20:

A student at a university is curious whether different majors have different GPAs. She knows that GPAs are normally distributed. She obtains random samples of 12 students from the psychology, physics, and English departments. She finds that the psychology students have a mean GPA of 3.44 (sd = 0.37), physics students have a GPA of 3.17 (sd = 0.36), and English students have a mean GPA of 3.72 (sd = 0.25).

15. What are the null and alternative hypotheses?

16. Have all assumptions been met to perform this test?

17. The student calculates that the variance SS between groups is 1.855, and the total SS is 5.47. Complete the table and calculate the appropriate test statistic.

Source	SS	df	s^2	F
Between				
Within				
Total				

18. What is your statistical conclusion?

19. Perform all pairwise post-hoc comparisons.

20. Interpret the post-hoc comparisons.

Chapter Ten Study Guide Answers

1. d

2. g

3. b

4. b

5. d

6. e

7. d

8–12. **ANOVA**

Source	SS	df	s^2	F_{obt}	F_{crit}
Between	1253.68	3	417.89	3.9980	2.84
Within	3762.72	36	104.52		
Total	5016.40	39			

8. Fill in the missing values.

$SS_W = SS_T - SS_B = 5016.40 - 1253.68 = 3762.72$

$df = df_T - df_B = 39 - 3 = 36$

$S_B^2 = SS_B / df_B = 1253.68 / 3 = 417.89$

$S_W^2 = SS_W / df_W = 3762.72 / 36 = 104.52$

$F_{obt} = S_B^2 / S_W^2 = 417.89 / 104.52 = 3.9981$

$F_{crit}(df = 3,36, \alpha = .05) = 2.84$

Reject the H_0.

9. 4
10. 10
11. 2.84
12. Yes

13. $\overline{X}_1 = 13.5000$ $\overline{X}_2 = 14.000$ $\overline{X}_3 = 13.6667$ $\overline{X}_4 = 8.8333$
$S_1 = 2.4290$ $S_2 = 2.7568$ $S_3 = 2.6583$ $S_3 = 2.4833$
$S_1^2 = 5.9000$ $S_2^2 = 7.6000$ $S_3^2 = 7.0667$ $S_3^2 = 6.1667$

HOV $7.6000 / 5.9000 = 1.29 < 4$ Okay!

Source	SS	df	S^2	F_{obt}	F_{crit}
Between	108.40	3	36.11	5.40	3.10
Within	133.60	20	6.68		
Total	242.00	23			

$df_B = k - 1 = 4-1 = 3$ $df_W = N - k = 24-4 = 2$ 0 $df_T = N - 1 = 24-1 = 23$

$SS_W = S_W^2 (df_W) = 6.68 (20) = 133.60$
$SS_B = SS_T - SS_W = 242.00 - 133.60 = 108.40$
$S_B^2 = SS_B / df_B = 108.40 / 3 = 36.13$
$F_{obt} = S_B^2 / S_W^2 = 36.11 / 6.68 = 5.40$
$F_{crit}(3,20) = 3.10$

Reject H_0: Age does affect memory ability.

14. Tukey's HSD Test

* Rank groups smallest to largest.

	4	1	3	2
$\overline{X}$	8.8333	13.5000	13.6667	14.000
$\overline{\overline{X}} - \overline{X}$		4.6667	4.8334	5.1667
			0.1667	0.5000
				0.3333
Q_{obt}		4.4230	4.5810	4.8969
			0.1580	0.4739
				0.3159

Q_{crit} $= 3.90$
$\alpha = .05$, df $= 20$, k $= 4$

$df_B = k-1 = 4-1 = 3$ $df_W = N-k = 24-4 = 20$ $df_T = df_B + df_W = 23$

$Q_{obt} = \overline{X} - \overline{X} / \sqrt{(S_W^2 / n)}$

(denominator) $\sqrt{(S_W^2 / n)} = \sqrt{(6.68 / 6)} = 1.0551$

Conclusions: There was a significant difference between Group 4 (60 Years) and all three other groups (30, 40, and 50 Years). There were not significant differences between Groups 1, 2, and 3 (30, 40, and 50 Years).

15. H_0 = There is no difference between mean GPAs for the three majors.
H_A = There is at least one difference between the mean GPAs for the three majors.

16. Yes—GPA = interval/ratio data, random samples, GPA is normally distributed, HOV assumptions are met. HOV = $(0.37)^2/(0.25)^2 = 2.19$, $2.19 < 4$

17.

Source	SS	df	s^2	F
Between	1.855	3-1 = 2	1.855/2 = 0.9275	0.9275/0.1095= 8.47
Within	5.47-1.855 = 3.615	36-3 = 33	3.616/33 = 0.1095	
Total	5.47	36-1 = 35		

18. Fcrit (alpha = 0.05, df = 2, 33) = 3.29
$8.47 > 3.29$, reject H0.
There is a difference in the average GPAs of psychology, physics, and English majors.

19. Use Tukey's HSD:
Qcrit (alpha = 0.05, df = 2, 33) = 2.89

Psychology vs. Physics: $((3.44-3.17)/(\sqrt{0.1095/12})) = 2.83$
Accept H_0
Psychology vs. English: $((3.44-3.72)/(\sqrt{0.1095/12})) = -2.93$
Reject H_0
Physics vs. English: $((3.17-3.72)/(\sqrt{0.1095/12})) = -5.76$
Reject H_0

20. There is no significant difference between the GPAs of psychology and physics majors. However, there is a significant difference between psychology and English majors, and between physics and English majors. It appears that English majors have higher average GPAs than the other students.

Chapter Eleven

Study Guide Questions

Notes to Students

We told you there would be more tests, right? Fortunately, the tests that are added in this chapter are all variations of the ANOVA test that you learned about in the previous chapter. A Two-Way ANOVA is the same as a One-Way ANOVA that you learned in Chapter 8 except that you have two independent variables instead of one. A Multifactorial ANOVA is an ANOVA with more than two independent variables. A Repeated Measures ANOVA is an ANOVA where the design is within-groups rather than between-groups. These tests are common when studying change over time or age-related changes in behavior in animals or humans.

The good news is that you won't have to do much math on these tests. These tests are best done with a computer since the calculations can become cumbersome with multiple variables. You'll want to make sure you understand the differences between the tests and how to read a story problem to pick out how many independent variables and conditions there are in the study. Just like your work in Chapter 7, you'll also need to recognize when the story problem is describing a between-groups design versus a within-groups design. Once you've practiced this, you'll be able to choose the appropriate test correctly and master the chapter.

Conceptual Questions

Multiple Choice Questions

1. A "within-groups effect" refers to the effect of:

a. The independent variable(s) on the dependent variable
b. The independent variable(s) on the experimental groups
c. Having repeated observations on the same (or similar) individuals (ie. non-independent data points)
d. The independent variable(s) on the group means

2. In two-way ANOVA, the interaction effect measures the:

a. Effect of any interaction among the independent variable(s)
b. Interaction of the dependent and independent variable(s)
c. Degree to which the dependent variable is affected by observer bias
d. Effect of an interaction among the independent variables over and above the main effects

3. Multifactorial ANOVA is simply an extension of the linear model for:

a. One-way ANOVA
b. Two-way ANOVA
c. Repeated-measures ANOVA
d. Independent t-test

Short Answer Questions

4. In a research design that includes two main effects or factors, the proper technique would be two-way ANOVA. What information does a two-way ANOVA provide, over and above the information that you would obtain from performing two one-way ANOVAs?

5. Describe the difference between a between-groups effect and a within groups effect?

6. How is multifactorial ANOVA an extension of two-way ANOVA?

7. What makes a repeated measures one-way ANOVA more powerful than independent one-way ANOVA?

Define the following terms or symbols:

8. s_R^2

9. s_{RC}^2

10. Score $= \mu + IV_1 + IV_2 + IV_3 + (IV_1)(IV_2) + (IV_1)(IV_3) + (IV_2)(IV_3) + (IV_1)(IV_2)(IV_3)$

$+ E$

Combined (Concepts and Calculation) Questions

Use the information below to answer questions 11 and 12:

In a study of the perception of desirability in companions, a researcher designed 24 stimulus images in which specific parameters of facial appearance in a limited number of models were manipulated using digital technology. The three manipulations were in Symmetry, Eyebrow position, and Chin structure. Twelve stimuli images were male and 12 were female. The 24 images were presented mixed into a set of 62 additional dummy images. Eighteen male and 18 female subjects were presented with the images and asked to score each on "desirability" of a coffee break "date." The data measured the number of "desirable" choices made by each subject.

11. Complete the ANOVA source table below:

ANOVA

Source of Variation	SS	df	MS	F	F crit
Rows (Sex)	117.3611	1			
Columns (Facial characteristic)	867.5556	2			
Interaction (Sex X Facial)	57.5556	2			
Within	124.5000	30			
Total					

12. Interpret the outcome of the study based on the hypothesis and whether you can reject the null hypothesis for each F value.

Use the information below to answer questions 13 and 14:

It has been well-known for years that room size and room color can affect mood and anxiety in humans and animals. Now, through the use of functional magnetic resonance imaging (fMRI), we can actually measure the activity of specific parts of the brain in reaction to stimuli being presented to the subject. In this experiment, subjects were placed into an fMRI machine and were presented with images of rooms of three sizes (small, medium, and large) and four colors (red, yellow, green, and blue) through the use of virtual reality technology. In earlier tests, subjects performed the same on tests of response to room size and color in actual and virtual reality presentations of the stimuli. The data are measures of brain activity in specific regions of the brain.

13. Complete the ANOVA source table below:

ANOVA

Source of Variation	SS	df	MS	F	F crit
Rows (Room size)	460.0556	2			
Columns (Room color)	31141.1944	3			
Interaction (Size X Color)	3499.7222	6			
Within	7489.3333	24			
Total					

14. Interpret the outcome of the study based on the hypothesis and whether you can reject the null hypothesis for each F value.

Use the following information to answer questions 15 and 16:

A researcher wished to investigate if her treatment for obsessive-compulsive disorder (OCD) was efficacious. She was also interested in the role of monetary compensation on participant improvement, and whether there was an interaction between her treatment and subject payment. There were two treatment groups (treatment vs. control) and three different participant compensation amounts ($5, $10, $50). For her outcomes variable, she used the number of hours spent engaging in ritualized behavior. The researcher had a sample of 60 individuals diagnosed with OCD and logged their behaviors in a 24-hour period. Each participant is in only one condition, and there were an equal number of participants in each condition.

15. Complete the ANOVA source table below:

ANOVA

Source of Variation	SS	df	MS	F	F crit
Rows (Treatment)	240	1			
Columns (Compensation)	10	2			
Interaction (Treatment x Compension)	40	2			
Within	210.6	54			
Total	500.6	59			

16. Interpret the outcome of the study based on the hypothesis and whether you can reject the null hypothesis for each F value.

Use the following information to answer questions 17–19:

Researchers were interested in examining whether pairing of an electrical shock with a feared stimulus would impact retreat time in rats. They were interested in examining both location of the shock and whether timing impacted the pairing of the shock with the feared stimulus. The rats were implanted with an electrode in one of three brain areas (a neutral area, memory area A, or memory area B), and given an electrical impulse at a certain time interval after crossing a predetermined line and being presented with a fear-inducing stimulus (50 milliseconds, 100 milliseconds, or 150 milliseconds). Retreat time was recorded to determine if rats showed significant change after only one presentation of the feared stimulus.

17. State all possible null and alternative hypotheses:

18. Complete the ANOVA source table below:

ANOVA

Source of Variation	SS	df	MS	F	F crit
Rows (Site)	321.5295	2			
Columns (Time)	169.45	2			
Interaction (Site x Time)	446.93	4			
Within	371.96	36			
Total	1971.77	44			

19. Interpret the outcome of the study based on the hypothesis and whether you can reject the null hypothesis for each F value.

Chapter Eleven Study Guide Answers

1. c

2. d

3. b

4. Performing a two-way ANOVA instead of 2 one-way ANOVAs provides a test of significance for the interactive effect of the independent variables on the dependent variable (in addition to reducing the chances of Type I error, of course!).

5. A between-groups effect is an effect of the independent variable on the mean of the treatment groups while a within-groups effect is the variability or "effect" of individual variability in response to the independent variable; it is a measure of variability "within the group."

6. Multifactorial ANOVA is an extension of two-way ANOVA in that it simply adds additional main effect terms and interaction terms to the linear model. A two-way ANOVA is an extension of one-way ANOVA in adding the interaction term to the linear model.

7. A repeated measures one-way ANOVA is more powerful than an independent one-way ANOVA because it removes more of the unexplained variance. Specifically, it removes the portion of the variance that was due to individual or subject differences. Having a smaller denominator (because we have reduced the unexplained error) results in a larger F-obtained value, making it more likely that we can reject the null hypothesis when it is actually false.

8. Between-row variance (reflects natural variation in the population and any variability due to your row variable)

9. Row × Column variance (reflects natural variation in the population and any variability due to the interaction between your row and column variables)

10. This is an equation for a multifactorial ANOVA, where
μ = the grand mean,
IV_1 = effect of the first independent variable,
IV_2 = effect of the second independent variable,
IV_3 = effect of the third independent variable,
$(IV_1)(IV_2)$ = interaction effect of the first two independent variables,
$(IV_1)(IV_3)$ = interaction effect of the first and third independent variables,
$(IV_2)(IV_3)$ = interaction effect of the second and third independent variables,
$(IV_1)(IV_2)(IV_3)$ = interaction effect of all three independent variables,
E = unexplained error.

11 and 12:

ANOVA

Source of Variation	SS	df	MS	F	F crit
Rows (Sex)	117.3611	1	117.3611	28.2798	4.1709
Columns (Facial characteristic)	867.5556	2	433.7778	104.5248	3.3158
Interaction (Sex X Facial)	57.5556	2	28.7778	6.9344	3.3158
Within	124.5000	30	4.1500		
Total	1166.9722	35			

The main and interaction effects are all significant in this analysis: there was a significant difference between males and females as to their average score (number of desirable images), there was a significant effect of facial characteristic on the average score awarded by the subjects, and there was a significant interaction between sex and facial characteristic in which males and females differed in their choice of most desirable facial characteristic modifications.

13 and 14:

ANOVA

Source of Variation	SS	df	MS	F	F crit
Rows (Room size)	460.0556	2	230.0278	0.7371	3.4028
Columns (Room color)	31141.1944	3	10380.3981	33.2646	3.0088
Interaction (Size X Color)	3499.7222	6	583.2870	1.8692	2.5082
Within	7489.3333	24	312.0556		
Total	42590.3056	35			

The room size effect was not significant, thus we detected no difference in the brain activity under conditions of different room size (at least at this brain location). Room color did produce a significant effect; brain activity at this location differed among the room color treatments. There was no significant interaction effect; the pattern of room color effects was the same across each room size treatment.

15 and 16:

ANOVA

Source of Variation	SS	df	MS	F	F crit
Rows (Treatment)	240	1	240	6.53	4.03
Columns (Compensation)	10	2	5	1.28	3.18
Interaction (Treatment x Compensation)	40	2	20	5.13	3.18
Within	210.6	54	3.9		
Total	500.6	59			

The treatment condition was significant, thus we detected a significant difference in the number of hours spent performing ritualized behavior based on whether the subject received the treatment or not.

Compensation did not produce a significant effect; hours of ritualized behavior were not significantly different based on amount subjects were being paid. There was also a significant interaction effect; the pattern of treatment effect varied across the different levels of compensation.

17–19:
Hypotheses are:

Null hypothesis (main effect for site): There is no difference in the retreat time based on the placement of the electrode, or the differences are due to chance.
Alternative hypothesis (main effect for site): There is a difference in retreat time based on the placement of the electrode.
Null hypothesis (main effect for time): There is no difference in retreat time based on time of presentation of the feared stimulus, or the differences are due to chance.
Alternative hypothesis (main effect for time): There is a difference in retreat time based on time of presentation of the feared stimulus.
Null hypothesis (interaction: site by time): There is no difference in retreat time based on the interaction between placement of the electrode and time of presentation of the feared stimulus over and above the differences that are due to the main effects of electrode placement and time of presentation of the feared stimulus, or the differences are due to chance.
Alternative hypothesis (interaction: site by time): There is a difference in retreat time due to the interaction between placement of the electrode and time of presentation of the feared stimulus.

ANOVA

Source of Variation	SS	df	MS	F	F crit
Rows (Site)	321.5295	2	160.79	15.57	3.28
Columns (Time)	169.45	2	84.725	8.21	3.28
Interaction (Site x Time)	446.93	4	111.7325	10.82	2.65
Within	371.96	36	10.32		
Total	1971.77	44			

Results show a significant effect of electrode placement site on retreat time. There was also a significant effect of timing as it related to fear stimulus presentation on rat retreat time. Finally, an interaction effect was observed between site of implantation and time of presentation of feared stimulus.

Chapter Twelve

Study Guide Questions

Notes to Students

This chapter might come earlier or later depending on your instructor's preferences for presenting this material, but we feel that this unit could be placed almost anywhere in the course because it stands alone quite well. This chapter is different from the other later chapters because it focuses on what analysis is appropriate when your independent variable is continuous rather than categorical, for example, if for your independent variable you are measuring and using actual intelligence scores (IQ) rather than grouping students into HIGH and LOW IQ.

The math requirements of this chapter will also likely vary according to the preferences of your instructor. None of the math is difficult, but there can be a lot of steps if you do it from the beginning (starting with the raw data). If your instructor is asking you to calculate regression and correlation from the raw data, you'll need to pay attention to the details in the sidebars of the chapter. This is where we explain how to calculate a regression and a correlation from scratch. Some instructors may instead have you do it from an intermediate point where they give you the sums of the raw data (as we demonstrate in the chapter). Either way you are asked to conduct the analyses, it will be important to perform the steps in the correct order. Recall that operations in parentheses are performed first and that you should not round at intermediate steps. Save the rounding for the final answer. There are a few conceptual points in this chapter that you'll need to learn, but the chapter as a whole does not seem to be particularly challenging on a conceptual level for most students.

Conceptual Questions

True or False Question
1. Correlation assumes that your data are homoscedastic.

Short Answer Questions
2. Name the 4 possible outcomes related to the results of a correlation.

3. If your standard error of the estimate (SEE) is large, what does this say about your predictions of Y?

4. You have just run a Pearson's r test on your data (where X = hours of sleep the night before an exam and Y = the exam score) and the result is r = .67. True or false: This indicates that 67% of the variability in exam scores is explained by the number of hours of sleep had the night before the exam.

5. Explain the difference between a correlation and a regression.

Combined (Concepts and Calculation) Questions

Use the following information to complete questions 6–10.

A researcher at the corporate office of Dreyer's has noticed that when ice cream sales go up there also appears to be a rash of crimes. To test whether there is a relationship between crime and ice cream, the researcher gathers a list of ice cream sales during the past 4 years (Note: ice cream sales are evaluated 4 times a year) and the corresponding crime rates.

Ice cream sales (millions $)	X	Crime rate (thousands)	Y		
Ice cream sales (Millions $) (X)	X^2	Crime rate (thousands) (Y)	Y^2		$X*Y$
8		125			
7.5		134			
12.3		203			
3.6		149			
6.7		210			
8.3		173			
13.4		340			
6.7		285			
7.4		198			
9.2		105			
17.8		207			
4.3		185			
5.9		165			
7.2		172			
15.3		323			
3.8		145			

6. Complete the table, solving for:
$\sum X$
$\sum X^2$
$\sum Y$
$\sum Y^2$
$\sum XY$

7. Calculate the Pearson r correlation value according to the data she collected.

8. What is the conclusion based on your results from your calculations?

9. What is the proportion of total variance of crime rate that can be explained by ice cream?

10. Graph the data points from above and draw the best fit line for the data.

11. Which of these two figures would have the strongest correlation?

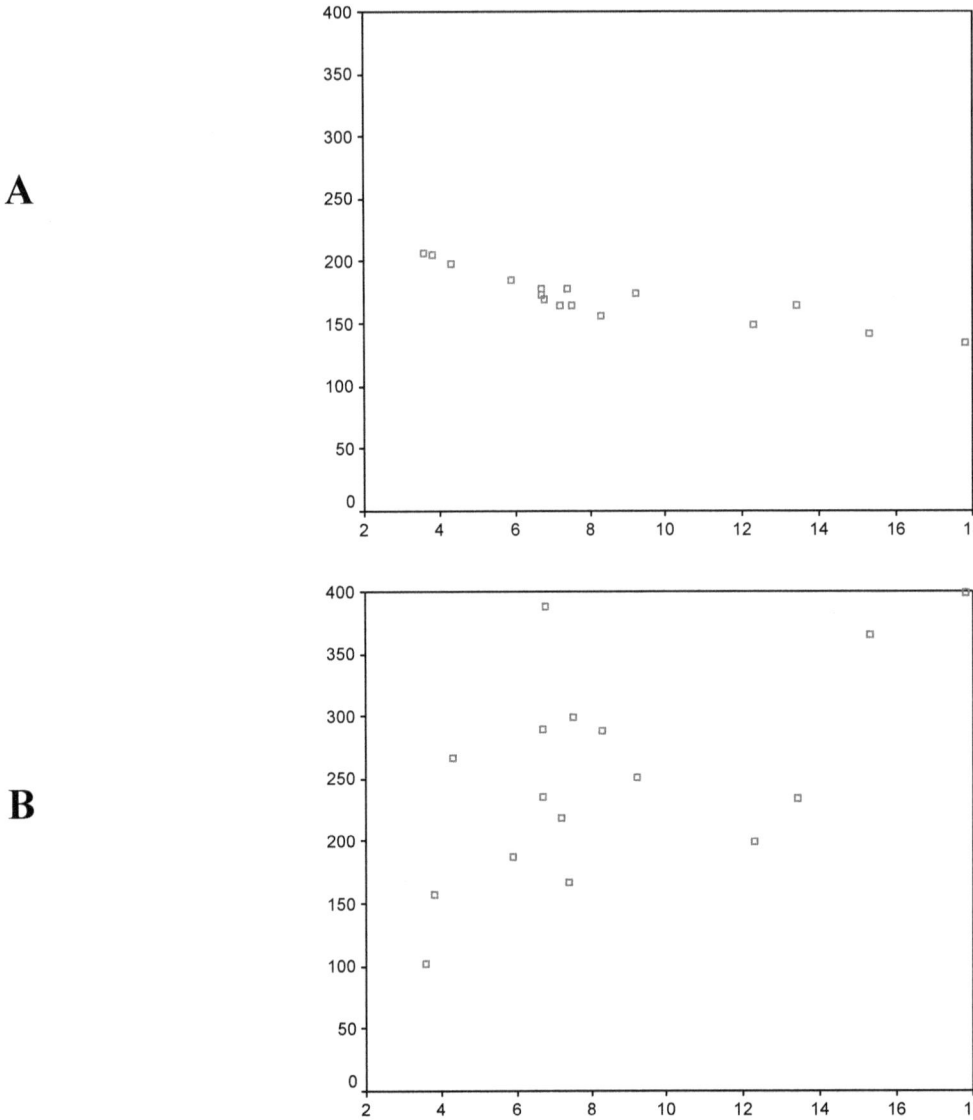

A

B

Use the following information to evaluate questions 12–16:

A neuroscientist suspects that low levels of serotonin may be causally related to aggressive behavior. She observes a group of 8 rhesus monkeys and records their serotonin levels and the number of aggressive acts performed. She runs a regression analysis on her data and obtains the following.

Y-intercept = 6.939
Slope = -7.127
r = -.75

12. Evaluate your results using rule of thumb.

13. Calculate r^2. What does this tell you?

14. Is there a significant relationship between serotonin levels and aggressive behavior? Use $\alpha = .05$.

15. How many aggressive acts would you predict to occur if a subject had a serotonin level of .49?

16. We observe a rhesus monkey in the wild and obtain a serotonin level of .3 and observe 6 aggressive acts. What would be the expected number of aggressive acts be at a serotonin level of .3?

Use the following information to answer questions 17–21.

A researcher is interested in studying the relationship between training and obedience in dogs. It is known that more training leads to better obedience. Data are collected from 23 dogs, and below is a graph of the relationship between number of training sessions and obedience scores.

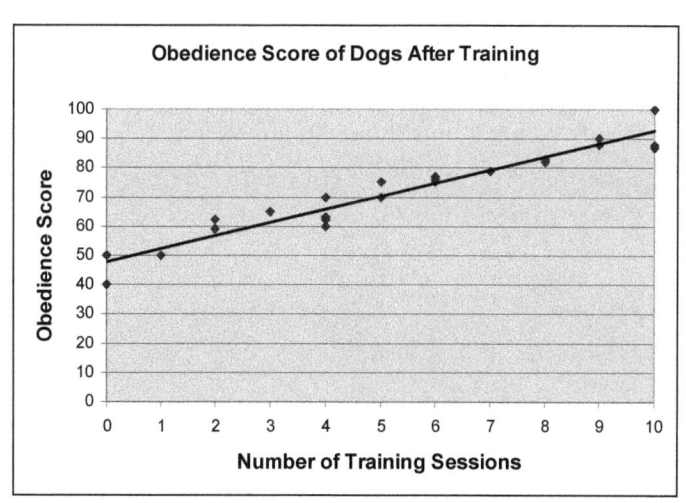

y-intercept = 47.711
slope = 4.5012
r = 0.9647

17. Evaluate the correlation using rule of thumb.

18. Calculate r^2. What does this tell you?

19. Is there a significant relationship between number of training sessions and obedience scores? Use $\alpha = .05$.

20. What is the equation for the line on the above graph? (Include actual numbers in your equation where appropriate.)

21. If a dog has 6 training sessions, what is his expected obedience score?

Chapter Twelve Study Guide Answers

1. False (but homoscedacity is an assumption of regression)

2.
- there is no relationship between X and Y
- X causes Y
- Y causes X
- There is a third variable influencing/correlated with both X and Y.

3. The larger the SEE, the less confident we can be in our predictions of Y.

4. False: you would need to calculate $R^2 = .4489$. Your results indicate that 44.89% of the variability in exam scores is explained by number of hours of sleep.

5. Regression focuses fitting the best line to a series of data points so that predictions of a Y-value can be calculated given an X-value. The primary component is the slope.

Correlation focuses on the strength (how tightly grouped a set of data points are) and the direction of those data points.

6.

Ice Cream Sales (Millions $) ($X$)	X^2	Crime rate (thousands) (Y)	Y^2	$X*Y$
8		125		
7.5		134		
12.3		203		
3.6		149		
6.7		210		
8.3		173		
13.4		340		
6.7		285		
7.4		198		
9.2		105		
17.8		207		
4.3		185		
5.9		165		
7.2		172		
15.3		323		
3.8		145		
$\sum X = 136.2$	$\sum X^2 = 1414.88$	$\sum Y = 3119$	$\sum Y^2 = 677311$	$\sum XY = 28812.8$

7. numerator = 2262.3125
Denominator = 4207.71081
$r = .5377$

8. There is a modest positive relationship between ice cream sales and crime rate. That is, as ice cream sales increase, so does crime rate.

9. $r^2 = 0.289$

10.

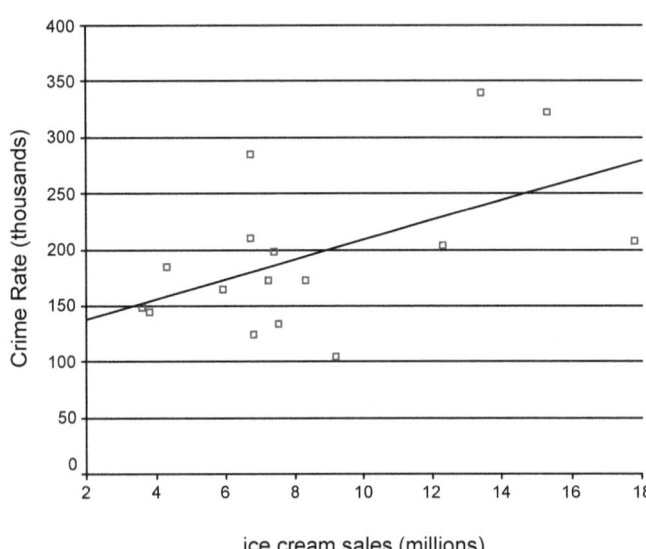

11. A.

12. -.75—Moderate correlation.

13. $r^2 = 56$. 56% of the variance in serotonin levels and aggressive behavior is due to the relationship between them.

14. R_{crit} (df = 8-2=6) = .707. | r_{obt} | › r_{crit} : Reject the null. There is a significant correlation between serotonin levels and aggressive behavior.

15. Y' = bx+a
Y' = -7.127(X)+.6939

Y' = -7.127(.49)+.6939= 3.45

16. $y' = -.5.82 (.3) + 7.54 = 5.794$

17. Strong correlation

18. $r^2 = .931$

19. R_{crit} (df = 23-2 = 21) = .413. | r_{obt} | › r_{crit} : Reject the null. There is a significant correlation between number of training sessions and obedience score.

20. Since y-intercep = 47.711 and slope = 4.5012, the equation is $Y = 47.711 + 4.5012X$.

21. $Y = 47.711 + 4.5012X$
    $Y = 47.711 + 4.5012 (6)$
    $Y = 74.7182$

You would expect an obedience score of 74.7812 after 6 training sessions.

Chapter Thirteen

Study Guide Questions

Notes to Students

This chapter might be dropped from some introductory courses but is a particularly important chapter for students who may be conducting statistics in their chosen profession. The material introduced here is conceptually an extension of the ANOVA and regression concepts that were addressed in previous chapters, but it illustrates how these tests are actually mathematically related to one another. Once you understand that, you can see how minor modifications of the mathematical model underlying the tests allow you to create the formulas for Two-Way ANOVA, Multifactorial ANOVA, Repeated Measures ANOVA, Multiple Regression, and a new test called Analysis of Covariance (ANCOVA).

Fortunately, all of these tests are best done with a computer, so you won't be responsible for hand-calculating any of the statistics addressed in this chapter. However, we do show you how a statistics software package is designed so that you select the "GENERAL LINEAR MODEL" option directly off the menu. Thus, this is a practical chapter as well as a conceptual chapter for those who may go on to advanced statistics.

Conceptual Questions

Multiple Choice Questions
1. If your data contain a series of measurements on the same subjects over time, an appropriate analysis technique would be:
 a. Multiple regression
 b. Analysis of covariance
 c. Three-way analysis of variance
 d. Repeated measures ANOVA
 e. General linear model analysis

2. Why is that the appropriate statistical test for question 1?

3. If you wish to eliminate the effect of a confounding variable in your analyses, an appropriate analysis technique would be:
 a. Multiple regression
 b. Analysis of covariance
 c. Three-way analysis of variance
 d. Repeated measures ANOVA
 e. General linear model analysis

4. Why is that the appropriate statistical test for question 3?

5. If you wish to tease apart the relative effects, or lack thereof, of three categorical independent variables and to examine the degree to which the three variables interact in their influence on an interval or ration-type dependent variable, an appropriate analysis technique would be:
   a. Multiple regression
   b. Analysis of covariance
   c. Three-way analysis of variance
   d. Repeated measures ANOVA
   e. General linear model analysis

6. Why is that the appropriate statistical test for question 5?

7. If you want to examine the relative effects of a categorical variable such as gender and a continuous variable such as number of years of schooling, and the interaction of these two variables, on a continuous dependent variable such as age of marriage, an appropriate analysis technique would be:
   a. Multiple regression
   b. Analysis of covariance
   c. Three-way analysis of variance
   d. Repeated measures ANOVA
   e. General linear model analysis

8. Why is that the appropriate statistical test for question 7?

Short Answer Questions
9. Under what conditions should Repeated Measures ANOVA be used?

10. What is the benefit of performing Repeated Measures ANOVA over, say, One-way ANOVA?

11. Under what conditions should Multiple Regression be used?

12. What is the benefit of performing Multiple Regression over, say, a series of simple regressions?

Chapter Thirteen Study Guide Answers

1. To be specific, "d," and more generally, "e."

2. Repeated measures ANOVA will give you the greatest power when examining the changes in a single subject over repeated exposures.

3. To be specific, "b," and more generally, "e."

4. ANCOVA will remove the variance of a 3rd, unknown variable that may be influencing the results seen in the dependent variable but is not a variable in the study or experiment of interest.

5. To be specific, "c," and more generally, "e."

6. A three-way ANOVA is most appropriate because you have three independent variables, and you want to examine the unique influence of each on your dependent variable.

7. e

8. GLM allows you to maintain the useful information of your continuous independent variable (without needing to categorize it, like you would for an ANOVA) in a comparison with your categorical independent variable. Regression or ANOVA alone would not allow for this comparison.

9. Repeated measures ANOVA should be used when your data contain repeated measurements on the same subjects, or on subjects that have been specifically matched for certain characteristics. RM ANOVA is the equivalent of a paired t-test when you have more than two treatment groups.

10. Using RM ANOVA allows you to account for a greater amount of the unexplained error in your analysis by estimating the amount of error that is due to variation among individual subjects, and in so doing, increases the power of your analysis.

11. Multiple regression should be used when you have two or more continuous independent variables and a continuous dependent variable.

12. Performing a single multiple regression decreases the problem of Type I error rate inflation, just as performing an ANOVA is better than performing many independent t-tests.

Chapter Fourteen

Study Guide Questions

Notes to Students

The most difficult aspect of this chapter is not the math so much as the sheer volume of new tests that are presented. Frankly, despite this chapter's placement near the end of the textbook, the math is even easier than the math presented in some of the earlier chapters. However, the number of tests can overwhelm a statistics student, particularly early in the course. Keep in mind how far you have come and that this chapter is really just an extension of what you have already done!

Except for the Chi-square test, the other new tests are the non-parametric test equivalent of tests you've already learned. For example, the Wilcoxon test is the non-parametric version of a Paired t-test, the Mann-Whitney U test is the non-parametric version of the Independent t-test, and the Kruskal-Wallis test is the non-parametric equivalent of a one-way ANOVA. That means that you won't have to spend too much conceptual energy on understanding the experimental designs for these tests; you'll just have an option for what test to perform when you do not meet the assumptions of your already-familiar parametric tests.

Your challenge with this material will be to become even more proficient at working with critical value tables at the end of the book and at choosing the appropriate test given a story problem. Pay close attention to the rules for determining whether or not the test is significant because the Mann-Whitney U test requires that your obtained value is LESS THAN your critical value to be significant. A lower U obtained value actually means that there is less overlap between your two conditions (suggesting a bigger difference between your groups), so it does make sense once you realize what the statistic is calculating.

The story problems now need to be evaluated based on whether or not you meet the assumptions for the parametric tests, as well as the experimental design and the number of groups/conditions. If you have not used the flowchart so far, it might be worth another look now that there are so many tests to choose from. It can help you ask yourself all of the important questions so that the appropriate test is revealed from your answers to the questions about the design, etc.

Combined (Concepts and Calculation) Questions

Use the following information for questions 1–3.

A researcher believes that individuals in different occupations will show differences in their ability to be hypnotized. Six lawyers, six physicians, and six professional dancers are randomly selected for the experiment. A test of hypnotic susceptibility is administered to each. The results are shown below. The higher the score, the higher the hypnotizability.

Lawyers	Physicians	Dancers
26	14	30
17	19	21
27	28	35
32	22	29
20	25	37
25	15	34

1. What is the appropriate statistical test?

2. Conduct the appropriate statistical test, using $\alpha = .05$.

3. What can you conclude? If you reject the null, conduct any appropriate post-hoc analyses.

Use the following information to complete questions 4–6.

An investigator believes that students who rank high in certain kinds of motives will behave differently in gambling situations. To investigate this hypothesis, the investigator randomly samples 50 students high in affiliation motivation, 50 students high in achievement motivation, and 50 students high in power motivation. The students are asked to play the game of roulette, and a record is kept of the bets they make. The data are then grouped into the number of subjects with each kind of motivation who make bets involving low, medium, and high risk. Low risk means that they make bets involving low odds, medium risk involves bets of medium odds, and high risk involves playing long shots. The following data are obtained:

	Type of Motivation			
	Affiliation	Achievement	Power	
Low risk	26	13	9	48
Medium risk	16	27	14	57
High risk	8	10	27	45
	50	50	50	150

4. What is the appropriate statistical test?

5. Conduct the appropriate statistical test, using an α = .05.

6. What can you conclude—is there a relationship between these different kinds of motives and gambling behavior?

Use the following information to answer questions 7–9:

Since muscle tension in the head region has been associated with tension headaches, you reason that if the muscle tension could be reduced, perhaps the headaches would decrease. You design an experiment in which nine subjects with tension headaches participate. The subjects keep daily logs of the number of headaches they experience during a 2-week baseline period. Then you train them to lower their muscle tension using a biofeedback device. After 6 weeks of training, they again keep a 2-week log of the number of headaches experienced.

	# of headaches	
Subject	**Baseline**	**After Training**
1	17	3
2	13	7
3	6	2
4	5	3
5	5	6
6	10	2
7	8	1
8	6	0
9	7	2

7. What is the appropriate statistical test?

8. Conduct the appropriate statistical test, using α = .05, 2- tailed.

9. Given your calculations, what can you conclude?

Use the information below to answer questions 10–12:

The head of the marketing division of a leading soap manufacturer must decide among four differently styled wrappings for the soap. To provide data for the decision, he has the soap placed in the different wrapping styles and distributed to 5 supermarkets. At the end of 2 weeks, he finds that the following amounts of soap were sold:

Wrapping A	Wrapping B	Wrapping C	Wrapping D	
90	98	130	82	400

10. What is the appropriate statistical test?

11. Is there sufficient basis for making a decision about the wrappings? Use $\alpha = .05$.

12. What, if anything, can you conclude?

Use the following information to answer questions 13–15:

An ornithologist thinks that injections of follicle-stimulating hormone (FSH) increase the singing rate of his captive male cotingas (birds). To test this hypothesis, he randomly selects 20 singing contingas and divides them into two groups of 10 birds each. The first group receives injections of FSH and the second injections of saline solution, as a control for the trauma of an injection. He then records the singing rate (in songs per hour) for both groups. The results are provided below. Note that two of the FSH birds escaped during injection and were not replaced.

Saline	FSH
17	10
31	30
14	37
12	41
29	16
23	45
7	34
19	57
28	
3	

13. What is the appropriate statistical test?

14. Conduct the appropriate test using $\alpha = .05$, 1-tailed.

15. What can you conclude?

Use the following information to answer questions 16–18:

A developmental psychologist would like to determine whether infants display any color preferences. A stimulus consisting of four color patches (red, green, blue, and yellow) is projected onto the ceiling above a crib. Infants are placed in the crib, one at a time, and the psychologist records how much time an infant spends looking at each of the four colors. The color that receives the most attention during a 100-second test period is identified as the preferred color for that infant. The preferred colors for a sample of 60 infants are the following:

Red	Green	Blue	Yellow
21	11	18	10

16. What is the appropriate statistical test?

17. Conduct the appropriate test, using $\alpha = .05$.

18. What can you conclude? Do these dates indicate any significant preference among the four colors?

Use the following information to answer questions 19–21:

A university counselor believes that hypnosis is more effective than the standard treatment given to students who have high test anxiety. To test his belief, he randomly divides 22 students with high test anxiety into 2 groups: one receives the hypnosis treatment while the other receives the standard treatment. When the treatments are concluded, each student is given a test anxiety questionnaire. High scores indicate high anxiety.

Hypnosis	Standard
20	42
21	35
33	30
40	53
24	57
43	26
48	37
31	30
22	51
44	62
30	59

19. What is the appropriate statistical test?

20. Conduct the appropriate test, using α = .05.

21. What can you conclude from these data?

Use the following information to answer questions 22–24:

A sociologist is interested in whether there is a relationship between cohabitation before marriage and divorce. A random sample of 150 couples that were married in the past 10 years in a Midwestern city were asked if they lived together prior to getting married and if their marriage was still intact. The following results were obtained:

	Divorced	Still Married	
Cohabitated before marriage	58	42	100
Did not cohabitate before marriage	18	32	50
	76	74	150

22. What is the appropriate statistical test?

23. Conduct the appropriate test, using α = .05.

24. What can you conclude from these data?

Use the following information to answer questions 25–27:

You are interested in determining whether an experimental birth control pill has the side effect of changing blood pressure. You randomly select 10 women in Seattle. You give 5 of them a placebo for a month and then measure their diastolic blood pressure. Then you switch them to the birth control pill for a month and again measure their blood pressure. The other 5 women receive the same treatment except they are given the birth control pill first for a month, followed by the placebo for a month. The blood pressure readings are shown here.

A. Subject	Birth Control Pill	Placebo
1	108	102
2	76	68
3	69	66
4	78	71
5	74	76
6	85	80
7	79	82
8	78	79
9	80	78
10	81	85

25. What is the appropriate statistical test?

26. Conduct the appropriate statistical test, using $\alpha = .01$, 2-tailed.

27. What can you conclude?

Use the following information to answer questions 28–30:

A sleep researcher conducts an experiment to determine whether sleep loss affects the ability to maintain sustained attention. Fifteen individuals are randomly divided into the following three groups of 5 subjects each: group A, which gets the normal amount of sleep (7–8 hrs); group B, which is sleep-deprived for 24 hours; and group C, which is sleep-deprived for 48 hours. All three groups are tested on the same auditory vigilance task. Subjects are presented with half-second tones spaced at irregular intervals over a 1-hour duration. Occasionally, one of the tones is slightly shorter than the rest. The subject's task is to detect the shorter tones. The following percentages of correct detections were observed.

B. Normal Sleep	Sleep-Deprived: 24 hrs	Sleep-Deprived: 48 hrs
85	60	60
83	58	48
76	76	38
64	52	47
75	63	50

28. What is the appropriate statistical test?

29. Conduct the appropriate test, using $\alpha = .05$. Note: Determine *only* if there is an overall effect for sleep deprivation.

30. What can you conclude from these data?

Chapter Fourteen Study Guide Answers

1–3:
Kruskal-Wallis is the appropriate test.

C. Lawyers	R_1	Physicians	R_2	Dancers	R_3
26	10	14	1	30	14
17	3	19	4	21	6
27	11	28	12	35	17
32	15	22	7	29	13
20	5	25	8.5	37	18
25	8.5	15	2	34	16
	$\sum R_1 = 52,5$		$\sum R_2 = 34.5$		$\sum R_3 = 84$
	$\sum R_1{}^2 =$ 2756.25		$\sum R_2{}^2 =$ 1190.25		$\sum R_3{}^2 =$ 7056

$H_{OBT} = [12 / 18(19)][(2756.25/6) + (1190.25/6) + (7056/6)] - 3(18+1)$
    $= 7.365$

$H_{crit} = 5.991$; Reject the null hypothesis. People in different occupations do show differences in their abilities to be hypnotized.

4–6:
Chi-squared is the appropriate test.

f_e	$f_o - f_e$	$(f_o - f_e)^2$	$(f_o - f_e)^2 / f_e$
48(50)/150 = 16	26–16 = 10	100	100/16 = 6.25
57(50)/150 = 19	16–19 = -3	9	9/19 = 0.4737
45(50)/150 = 15	8–15 = -7	49	49/15 = 3.2667
48(50)/150 = 16	13–16 = -3	9	9/16 = 0.5625
57(50)/150 = 19	27–19 = 8	64	64/19 = 3.3684
45(50)/150 = 15	10–15 = -5	25	25/15 = 1.6667
48(50)/150 = 16	9–16 = -7	49	49/16 = 3.0625
57(50)/150 = 19	14–19 = -5	25	25/19 = 1.3158
45(50)/150 = 15	27–15 = 12	144	144/15 = 9.6
			$\sum = 29.57$

Chi-squared$_{obt}$ = 29.57
Chi-squared$_{crit}$ = 9.488; Reject the null hypothesis. There does appear to be a relationship between different kinds of motives (affiliation, achievement, and power) and gambling behavior.

7–9:

Wilcoxon Signed Rank Test is the appropriate test.

Baseline	After Training	D	Rank
17	3	14	9
13	7	6	5.5
6	2	4	3
5	3	2	2
5	6	-1	1
10	2	8	8
8	1	7	7
6	0	6	5.5
7	2	5	4

$\sum$**ranks (negative) = 1**
$\sum$**ranks (positive) = 45**

Tobt= 1
Tcrit = 6; Reject the null hypothesis. There does appear to be a difference in the number of tension headaches experienced when using the biofeedback device.

10–12:

Chi- Squared is the most appropriate test.

f_e	$f_o - f_e$	$(f_o - f_e)^2$	$(f_o - f_e)^2 / f_e$
400/4 = 100	90–100 = -10	100	1.00
400/4 = 100	98–100 = -2	4	0.04
400/4 = 100	130–100 = 30	900	9.00
400/4 = 100	82–100 = -11	324	3.24
			$\sum$=**13.28**

Chi-squared$_{obt}$ = 13.28
Chi-squared$_{crit}$ = 7.815; Reject the null hypothesis. The distribution of different wrapping styles is significantly different from random (and, for marketing purposes, he should choose Wrapping C).

13–15:

Mann-Whitney U is the most appropriate test.

Saline	R_1	FSH	R_2
17	7	10	3
31	13	30	11.5
14	5	37	15
12	4	41	16
30	11.5	16	6
23	9	45	17

7	2	34	14
19	8	57	18
28	10		
3	1		
$\sum R_1 = 70.5$		$\sum R_2 = 100,5$	

$U_1 = (10)(8) + [10(10+1)/2] - 70.5 = 64.5 \rightarrow$ U'obt
$U_2 = (10)(8) + [8(8+1)/2] - 100.5 = 15.5 \rightarrow$ Uobt
Ucrit = 20; Uobt < Ucrit ; Reject the null hypothesis. FSH does appear to increase the singing rate of captive male cotingas.

16–18:Chi-squared is the most appropriate test.

f_e	$f_o - f_e$	$(f_o - f_e)^2$	$(f_o - f_e)^2 / f_e$
60 / 4 = 15	21–15 = 6	36	2.4000
60 / 4 = 15	11–15 = -4	16	1.0667
60 / 4 = 15	18–15 = 3	9	0.6000
60 / 4 = 15	10–15 = -5	25	1.6667
			$\sum$=5.73

Chi-squared$_{obt}$ = 5.73
Chi-squared$_{crit}$ = 7.815; Fail to Reject the null hypothesis. The distribution of preferred colors is not significantly different from random.

19–21:
Mann-Whitney U is the most appropriate test.

Hypnosis	R_1	Standard	R_2
20	1	42	14
21	2	35	11
33	10	30	7
40	13	53	19
24	4	57	20
43	15	26	5
48	17	37	12
31	9	30	7
22	3	51	18
44	16	62	22
30	7	59	21
	$\sum R_1 = 97$		$\sum R_2 = 156$

$U_1 = (11((11) + [11(11+1)/2] - 97 = 90 \rightarrow$ U'obt
$U_2 = (11((11) + [11(11+1)/2] - 156 = 31 \rightarrow$ Uobt
Ucrit = 34; Uobt < Ucrit; Reject the null hypothesis. Hypnosis does appear to be more effective than the standard treatment for students who have high test anxiety.

22–25:
Chi-squared contingency table is the most appropriate test.

	Divorced	Still Married	
Cohabitated before marriage	58	42	100
Did not cohabitate before marriage	18	32	50
	76	74	150

f_e	$f_o - f_e$	$(f_o - f_e)^2$	$(f_o - f_e)^2 / f_e$
76(100)/150 = 50.6667	58 − 50.6667 = 7.3333	53.7773	53.7773/50.6667 = 1.0614
76(50)/150 = 25.3333	18 − 25.3333 = -7.3333	53.7773	53.7773/25.3333 = 2.1228
74(100)/150 = 49.3333	42 − 49.3333 = -7.3333	53.7773	53.7773/49.3333 = 1.0901
74(50)/150 = 24.6667	32 − 24.6667 = 7.3333	53.7773	53.7773/24.6667 = 2.1802

Σ**=6.45**

Chi-squared$_{obt}$ = 6.45
Chi-squared$_{crit}$ = 3.841; Reject the null hypothesis. There does appear to be a relationship between cohabitation before marriage and divorce.

25–27:
Wilcoxon Signed Rank Test is the most appropriate test.

Birth Control Pill	Placebo	D	Rank
108	102	6	8
76	68	8	10
69	66	3	4.5
78	71	7	9
74	76	-2	2.5
85	80	5	7
79	82	-3	4.5
78	79	-1	1
80	78	2	2.5
81	85	-4	6

Σ**ranks (negative) = 14**
Σ**ranks (positive) = 41**

Tobt = 14
Tcrit = 3; Fail to Reject the null hypothesis. The birth control pill does not appear to significantly affect blood pressure.

28–30:
Kruskal-Wallis is the most appropriate test.

Normal sleep	R_1	Sleep-Deprived: 24 hrs	R_2	Sleep-Deprived: 48 hrs	R_3
85	15	60	7.5	60	7.5
83	14	58	6	48	3
76	12.5	76	12.5	38	1
64	10	52	5	47	2
75	11	63	9	50	4
	$\sum R_1 = 62,5$		$\sum R_2 = 40$		$\sum R_3 = 17.5$
	$\sum R_1{}^2 = 3906.25$		$\sum R_2{}^2 = 1600$		$\sum R_3{}^2 = 306.25$

$H_{OBT} = [12 / 15(16)] [(3906.25/5) + (1600/5) + (306.25/5)] - 3(15+1)$
$= 10.125$

$H_{crit} = 5.991$; Reject the null hypothesis. Sleep loss does appear to affect the ability to maintain sustained attention.

Chapter Fifteen

Study Guide Questions

Notes to Students

In this chapter we review some important concepts for each of the previous chapters and we review how to choose the appropriate test. Choosing the appropriate test is the most challenging aspect of the review chapter. We have added another figure to aid you in achieving this skill, and that is a table that outlines the factors involved in choosing the appropriate test.

The story problems now need to be evaluated based on whether or not you meet the assumptions for the parametric tests, as well as the experimental design and the number of groups/conditions. If you have not used the flowchart so far, it might be worth another look now that there are so many tests to choose from. It can help you ask yourself all of the important questions so that the appropriate test is revealed from your answers to the questions about the design, etc. If you have tried to use the flowchart and have found that the flowchart has not helped you, then try looking at Table 13.2 to see if that helps. Some students prefer the table over the flowchart and you might be one of them!

If you find you are struggling with the exercises in the back of the chapter and/or the study guide problems, you may need to review information in earlier chapters. The most common problem our students have is distinguishing between nominal, ordinal, and interval/ratio measurement scales. You probably haven't thought too much about them since Chapter 2 and that might be causing a problem with your ability to pick the appropriate test since some tests require interval/ratio data, while others do not. If that is not the problem you are having, you may also be having difficulty picking out the information that you need from the story problem. Practice, practice, practice and underlining the important details in the paragraph (such as subjects "<u>were matched based on age and sex</u>") can help with this problem. Don't forget our earlier suggestion about seeking out additional examples from other sources (online or in textbooks at the library) as exposure to numerous problems seems to help our students with the evil story problem.

Choose-the-Appropriate-Test Questions

1. A clinical psychologist has noted that autistic children seem to respond to treatment better if they are in a familiar environment. To evaluate the influence of environment, the psychologist selects a group of 96 autistic children who are currently in treatment and randomly divides them into three groups. One group continues to receive treatment in the clinic as usual. The second group receives treatment sessions in the child's home. The third group gets half the treatments in the clinic and half at home. After 6 weeks, the psychologist evaluates the progress for each child. The variances for the three groups are 1.78, 4.05, and 3.16.

2. A physician employed by a large corporation believes that due to an increase in sedentary life in the past couple of decades, middle-aged men have become fatter. In 1970, the corporation measured the percentage of body fat in their employees and found the mean to be 22%. To test her hypothesis, the physician measures the fat percentage in a random sample of 50 middle-aged men currently employed by the corporation. The fat percentages are as follows: 24, 40, 29, 32, 33, 25, 15, 22, 18, 25, 16, 27.

3. A researcher suspects that increasing the level of lighting during the winter months will have a positive effect on people's moods. To test this hypothesis, she identifies a sample of 36 college students living on one floor of a dormitory. Each student is given a mood questionnaire that uses a 6-point scale at the beginning of February, and then all the lights on the dormitory floor are changed from 75-watt to 100-watt bulbs. After 4 weeks, the students are again given mood questionnaires and the researcher records the amount of difference between the two measurements.

4. A researcher examines the short-term effects of a single treatment with acupuncture on cigarette smoking. Seventy-five subjects who smoke approximately one pack of cigarettes a day are assigned to one of two groups. One group receives the treatment with the acupuncture needles. The second group of subjects serves as a control group and is given relaxation instruction while on the examining table but no acupuncture treatment. The number of cigarettes smoked the next day is recorded for all subjects. Group one had a mean of 7 and a standard deviation of 2.1, and group two had a mean of 9 with a standard deviation of 3.3.

5. Studies have suggested a personality type is related to heart disease. Specifically, type A people are more prone to heart disease, while type B people are less likely to have it. An investigator wants to examine the relationship between personality type and heart disease. For a random sample of individuals personality type (A or B) is assessed with a standardized test. These individuals are then examined and categorized according to the type of disorder (heart disease, vascular disorder, hypertension, or none) they have.

6. The animal learning course in a college's psychology department requires that each student train a rat to perform certain behaviors. The student's grade is partially determined by the rat's performance. The instructor for the course has noticed that some students are very comfortable working with the rats and seem to be successful in training their rats. The instructor suspects that these students may have previous experience with pets that gives them an advantage in the class. To test this hypothesis, the instructor gives the entire class a questionnaire at the beginning of the course. One question determines whether or not each student currently has a pet of any type at home. Based on the responses to this question, the instructor divides the class into two groups (students with pets, n = 10; students without pets, n = 15) and compares the rats' learning scores for the two groups.

7. A common science fair project involves testing the effects of music on the growth of plants. For one of these projects, a sample of 24 newly sprouted bean plants is obtained. These plants are randomly assigned to four treatments: rock music, heavy metal, country, and classical. The dependent variable is the height of each plant after 2 weeks.

8. A researcher is interested in establishing whether type of background music affects the quality of ice skating performances. A random sample of 30 skaters is selected from among the students of a large skating club. The skaters are randomly assigned to one of three conditions: A, jazz music; B, classical music; C, no music. Each skater performs for 10 minutes and the number of falls per skater is recorded; variances are homogenous.

9. A random sample of 122 delinquent boys was selected and randomly divided into two groups. The researcher was interested in discovering whether a 6-week, nondirective, individual therapy program would affect levels of anxiety measured on a scale of 1–10. The boys in group A all received the therapy, whereas those in Group B did not. Both groups were then given the anxiety level test; variances were homogenous.

10. A developmental psychologist was interested in the effect of age on self-control. She conducted a test of inhibition (or self-control) on 50 children when they were three years old. The test recorded the number of times children failed to inhibit their actions when they were told to do so by the experimenter. She retested the children when they were five years old and compared their results. The standard deviations were 1.3 and 4.9, respectively.

11. There are a number of different memory strategies that you can employ to improve the likelihood of remembering a fact or name. To test the efficiency of three strategies, researchers recruited 60 subjects and randomly assigned them to one of three groups (elaboration, rehearsal, or visualization). They measured the subject's recall of words after utilizing the memory strategy. The variances were homogeneous.

12. A marital therapist asked 46 husbands to rank their wives on their willingness to communicate about the couple's relationship on a scale from 1 to 10, and compared these scores to 30 husbands who were not in marital therapy. Was there a significant difference between the groups?

13. Oakes and colleagues explored the effect of race on identifying an individual as a criminal or noncriminal. They presented images of black and white individuals that were or were not holding a weapon. Thirty subjects were told to pretend they were a police officer and to "shoot" criminals with weapons (via mouse click) in the virtual simulation. Among other things, they measured the number of times the same subjects "shot" a person that was not holding a weapon when they were black versus when they were white, and found they were more likely to shoot incorrectly when the image was of a black person. The standard deviation was 2.2 and 8.1, respectively.

14. Many people are afraid of snakes and one possible treatment is to use a technique called flooding. A therapist compared the heart rates (heartbeats per minute) of 30 subjects that were treated with the "flooding" technique versus 45 who were not treated to measure their reaction when they were presented with a live (nonpoisonous) snake for 60 seconds. The mean and standard deviation for the flooding group was 78.6 and 1.4, respectively, while the mean and standard deviation for the no-treatment group was 90.7 and 1.9, respectively. Was there a difference between the groups?

15. A researcher was interested in establishing whether attendance in a preschool affects social maturity level of children. A random sample of kindergarten children was selected and watched closely by trained observers for one full week. The children were then rank-ordered on the basis of perceived social maturity. The children were then divided into two groups on the basis of whether or not they had previously attended a day-care center. Did preschool affect social maturity scores?

16. An education specialist is interested in whether coaching can have any effect on math SAT scores. A group of 100 high school seniors was randomly selected from a large metropolitan school district. The group was then randomly divided into two equally sized subgroups. One group was given 3 months of daily coaching in those math skills deemed important to the SAT, while the other group spent the same amount of time each day watching reruns of the TV show *Happy Days*. At the end of the 3-month period, all students took the SAT and their math scores were compared.

17. Some researchers have supported that because of academic and other frustrations, adolescents labeled as LD (learning disabled) would have more symptoms of depression and even possibly higher levels of suicide ideation than would non-LD adolescents. Two groups of 16-year-old students, one labeled LD and the other non-LD (50 male adolescents in each group), were selected on the basis of a certain school district's records. All students were then given the SIQ-JR (questionnaire) and the scores are ranked.

18. A nursing researcher wished to test the hypothesis that male business majors earn more in later life than do either male liberal arts or education majors. A random sample of alumni was selected from the university files from each of the 3 subject major categories. To attempt to control for length of experience on the job, all subjects were selected form the same graduating class—the class that graduated 10 years ago. All the selected alumni were contacted and asked to indicate their yearly incomes. Because a few of the subjects reported enormously high incomes, the resulting distribution became so skewed that it was decided to rank-order the incomes.

19. A graduate student in nursing wanted to find out whether IQ is a function of family size. The speculation was that in families with fewer children, each child receives more parental attention and intellectual stimulation and should therefore have a higher IQ than would a child reared in a large family. A large random sample (n = 150) of two-child families was selected as well as a sample (n = 130) of six-child families. The IQs of all children were measured, variances are homogenous, and the two sample groups were compared.

20. An investigator tried to shed light on the hypothesis that perception shapes attitudes. A large random sample was selected, and each subject was then randomly assigned to one of three groups. Each group then heard an identical speech, given by the same speakers, on the topic of land reform in Cuba. To Group A, the speaker was introduced as a prominent political scientist; to Group B, as a member of the US State Department; and to Group C, as a member of the Cuban delegation to the UN.

After the speech, the subjects all took an "Attitude toward Cuba" test that measured reaction times to positive images about Cuba presented on a computer screen. The distribution of test scores was normal and the variances between groups are considered equal.

21. A grocery store executive had received complaints from customers that their 10 lb bags of Russet potatoes were smaller than normal. Because the store had recently been using a new supplier, the executive asked a consultant to conduct a study of the potatoes received in the last week and determine if they were being shorted. The consultant measured the weights of 550 bags of potatoes and found that the average weight was 9.67 lbs with a standard deviation of 0.50 lbs. The standard deviation for the population of bags of 10 lb potatoes is 1.4. What is the appropriate analysis to determine if the weights are significantly different?

22. A physical education professor believes that exercise can slow down the aging process. For the past 10 years, he has been conducting an exercise class for 14 individuals who are currently 50 years old. Normally, as one ages, maximum oxygen consumption decreases. The national norm for maximum oxygen consumption in 50-year-old individuals is 30 milliliters per kilogram per minute. Oxygen consumption is normally distributed. The mean of the 14 individuals is 40 milliliters per kilogram per minute.

Chapter Fifteen Study Guide Answers

1. ANOVA

2. SINGLE-SAMPLE t-test

3. WILCOXON (ordinal data)

4. INDEPENDENT T-TEST

5. CHI-SQUARED TEST

6. MANN-WHITNEY U

7. KRUSKAL-WALLIS (normality violated)

8. KRUSKAL-WALLIS

9. MANN-WHITNEY U

10. PAIRED T-TEST

11. KRUSKAL-WALLIS

12. MANN-WHITNEY U

13. PAIRED T-TEST

14. INDEPENDENT T-TEST

15. MANN-WHITNEY U

16. INDEPENDENT T-TEST assuming that there is no serious violation to the HOV; MANN-WHITNEY U if there is a serious violation with regards to the HOV

17. MANN-WHITNEY U

18. KRUSKAL-WALLIS

19. INDEPENDENT T-TEST

20. ONE-WAY ANOVA

21. SINGLE SAMPLE Z-TEST

22. SINGLE SAMPLE T-TEST

SAGE Research Methods Online

The essential tool for researchers

**Sign up now at
www.sagepub.com/srmo
for more information.**

An expert research tool

- An **expertly designed taxonomy** with more than 1,400 unique terms for social and behavioral science research methods

- **Visual and hierarchical search tools** to help you discover material and link to related methods

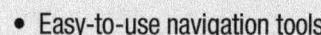

- Easy-to-use navigation tools
- Content organized by complexity
- Tools for citing, printing, and downloading content with ease
- Regularly updated content and features

A wealth of essential content

- The most comprehensive picture of quantitative, qualitative, and mixed methods available today

- More than **100,000 pages of SAGE book and reference material** on research methods as well as editorially selected material from SAGE journals

- More than **600 books** available in their entirety online

Launching 2011!

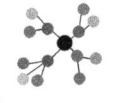

SAGE research methods online

In compliance with GPSR, should you have any concerns about the safety of this product, please advise: International Associates Auditing & Certification Limited The Black Church, St Mary's Place, Dublin 7, D07 P4AX Ireland EUAR@ie.ia-net.com